Skins, Envel

Integrate the best building envelope construction methods, materials science, and structural principles in your work using this book as a resource to help you . . .

With more than seventy significant case studies located in North America, South America, Europe, and Asia from prehistory to the present, this book illuminates the theory and techniques of assembling exteriors. Six chapters organized by wall types, from hand-set monolithic walls to digitally fabricated curtain walls, each have a material focus section to help you understand their intrinsic properties so that you can decide which will best keep the weather out of your building.

Examples from the ancient world, including **the Pyramids** and **the Great Wall**, through a range of renowned modern architects, such as **Studio Gang Architects**, **Sauerbruch Hutton**, **Herzog & de Meuron**, and **Rafael Moneo**, illustrate how significant works in the history of architecture explored innovative use of materials – stone, brick, concrete, glass, and aluminum. Along the way, principles of construction from masonry and basic framing through ever more sophisticated envelope systems address classic problems presented by gravity, wind, rain, and sun with studies of lateral forces, building movements and materials that bridge the gaps in between them.

Mayine L. Yu is a senior associate at Davis Brody Bond Architects in New York and was an adjunct associate professor in the Graduate School of Architecture, Planning, and Preservation at Columbia University.

Skins, Envelopes, and Enclosures
Concepts for designing building exteriors

Mayine L. Yu

Routledge
Taylor & Francis Group
NEW YORK AND LONDON

First published 2014
by Routledge
711 Third Avenue, New York, NY 10017

Simultaneously published in the UK
by Routledge
2 Park Square, Milton Park, Abingdon, Oxon OX14 4RN

Routledge is an imprint of the Taylor & Francis Group, an informa business

© 2014 Taylor & Francis

The right of Mayine L. Yu to be identified as author of this work has been asserted by her in accordance with sections 77 and 78 of the Copyright, Designs and Patents Act 1988.

All rights reserved. No part of this book may be reprinted or reproduced or utilised in any form or by any electronic, mechanical, or other means, now known or hereafter invented, including photocopying and recording, or in any information storage or retrieval system, without permission in writing from the publishers.

Trademark notice: Product or corporate names may be trademarks or registered trademarks, and are used only for identification and explanation without intent to infringe.

Every effort has been made to contact and acknowledge copyright owners. The publishers would be grateful to hear from any copyright holder who is not acknowledged here and will undertake to rectify any errors or omissions in future printings or editions of the book.

Library of Congress Cataloging in Publication Data
Yu, Mayine.
Skins, envelopes, and enclosures: concepts for designing building exteriors / Mayine Yu.
pages cm
Includes bibliographical references and index.
1. Exterior walls. 2. Buildings--Protection. 3. Architectural design. I. Title.
NA2940.Y8 2013
721--dc23
2012048826

ISBN: 978-0-415-89978-9 (hbk)
ISBN: 978-0-415-89979-6 (pbk)
ISBN: 978-0-203-78529-4 (ebk)

Typeset in Avenir by
Servis Filmsetting Ltd, Stockport, Cheshire

Acquisition Editor: Wendy Fuller
Editorial Assistant: Laura Williamson
Production Editor: Alanna Donaldson

Printed by Bell & Bain Ltd, Glasgow

This book is dedicated to the memory of my mother

Contents

Preface and acknowledgements — ix

Introduction — 1

Part 1 Load-bearing construction

1. Monolithic bearing walls – masonry, stone, concrete — 5
   - Historical survey — 5
   - Concrete masonry units (CMU) — 33
   - Mortar — 36
   - Brick terminology — 36
   - Structural bonds — 37
   - Bearing wall: masonry — 37
   - Masonry construction technique — 39
   - Dimensioning — 39
   - Jointing — 39
   - Reinforcement — 40
   - Openings — 40
   - Expansion/contraction — 40
   - Insulation — 40
   - Flashing — 42
   - Bearing walls: stone and concrete — 42
   - Concrete — 49

2. Load-bearing cavity walls — 61
   - Weather and rainwater — 62

3. Light frame construction — 79
   - Balloon framing — 84
   - Platform framing — 86

Part 2 Non-load-bearing construction

4. Frame and infill construction — 115
   - Windows and openings — 134
   - Windows and operation — 144
   - Proto-unitization — 149

5. Cladding systems — 156
   - Precast concrete spandrel — 170

Rainscreen/double seal systems	176
Cladding materials	178
Location of insulation and vapor barrier	178
Detailing concerns	179
Spandrel systems	181

6. Curtain walls — 197

Glass architecture	197
Materials and historical development	198
Fireproofing	207
Height, movement, and the development of unitized systems	216
Wind pressures and lateral forces	221
Panelization	224
Pressure equalization and unitized mullion detailing	227
Environmental forces	228
Finishes and form	232

Notes	245
List of figure credits	246
Bibliography	255
Index	259

Preface and acknowledgements

This text evolved from lectures given at the Columbia University Graduate School of Architecture, Planning, and Preservation in the Introduction to Construction Technology course sequence. Inevitably biased by time, place, construction methods and design principles prevalent in the locality, region, and environment, the information is therefore rooted in the New York City context. Influenced by the biases of particular education and work experiences, it also represents an attempt to begin thinking about the technology of building construction in the context of architectural history and the society in which modern architecture evolved. The result is not definitive, by any means, but represents one attempt to provide a supplement to the classic texts and reference materials with a basis for construction design and detailing. The class, and therefore this text, offers a brief survey of many subjects that would each benefit from more depth and elaboration, but still may have value as an overview of concepts and image resources for the student and teacher of architectural construction principles.

The class as it evolved over the years came to rely heavily on the resources of Avery Library, inevitably, and classic texts and design manuals common to many introductory survey courses – most notably Ed Allen and Joseph Iano's *Fundamentals of Building Construction*, especially the earlier editions from which the course was taught; Ed Ford's *Details of Modern Architecture*, both volumes; and Frank Ching's beautiful books, including *Building Construction Illustrated*. These texts informed my own education and together touch upon aspects that the course attempted to transmit to students of architecture in the graduate program. Though no overview survey class can do sufficient justice to any one of the topics touched upon briefly in the text, I hope that the combination of images, illustrations, and words will convey some portion of the information shared in the sequences of lectures laid out by a structural engineer but always taught in conjunction with an architect. This overview brought in examples from milestone projects in the history of architecture in an attempt to link the introductory architectural history and theory studies occurring concurrently for most students in the first-year program.

The materials included in this volume are organized in a format somewhat unusual for architectural education and for the study of construction detailing, in an attempt to present engaging examples from the history of architecture and technological developments in the materials and construction; they touch upon social, economic, and political connections, while still conveying the principles and fundamentals of construction detailing. The class, as it was once taught, included drawing assignments, readings, introductory detailing tied to calculations for environmental performance and heat loss, as well as detail drawings and construction of a full- or half-scale model selected from the construction types detailed in chapters 2–6. Given the space constraints of the New York City campus, these were necessarily limited in size, though memories of my own first construction detailing class with bricklaying, stud framing, and concrete pouring in a supervised architectural lab remain vivid.

Gratitude is due to the illustrious predecessors who taught the course prior to me, including Thomas Hanrahan, Jackie Welsh, Areta Pawlynsky, and Lois Mate. They, like I, had the privilege of teaching the class alongside Anthony C. Webster, Director of the Building Technology sequence at the time. Thanks are also due to former employers – the late Robert Gutman at Princeton, Kenneth Frampton for a brief period at Columbia, and Robert A. Heintges, at his company – for the opportunities they offered that contributed so much to my education, research, and practical experiences. Special thanks, most of all, to Areta Pawlynsky for lessons great and small, while working together and ever since; and Mary McLeod for offering intellectual rigor, support and kindness through the years.

Mayine L. Yu
New York, September 2012

Introduction

Based on a Columbia University Graduate School of Architecture, Planning, and Preservation first-year course taught by a structural engineer in tandem with an architect, the contents of this book are rooted in a New York/North American point of view. While the intention is to apply the principles of design for exterior walls to a broad range of environments without cultural judgment, it is inevitable that some vestiges of the course remain and may puzzle some readers. As a complement to other text and reference manuals, the examples and illustrations shown strive to bridge the gap between technical realities of construction materials and the rich architectural heritage, historical meaning, and design possibilities of the palette of options available for building skins today.

The structural viewpoint originally established by the engineering professor logically sets us up to survey the overall categories of wall construction from their weight and load-bearing capacities, as a beginning point. We therefore examine load-bearing wall systems and construction classes before non-load-bearing exterior enclosures. This generally corresponds with a chronological and historical survey of architectural history, though examples from more recent periods are brought in to illustrate how each of these systems influenced the evolution of architectural construction. In some cases, the new construction technologies were adopted by architects and embedded within the design aesthetic and architectural statements they sought to establish.

Iconic examples of modern architecture throughout the history of architectural development (primarily Western) are therefore used wherever possible as illustrations of construction and technical innovation. In some examples, new technologies blinded architects to the traditional detailing techniques that had evolved in their local building traditions to address environmental realities. In other illustrative examples, architects and builders were able to bring the best of emerging technologies together with the traditions of construction using materials and assemblies to express a design aesthetic in ways that suggest design opportunities for the future.

Part 1 | Load-bearing construction

Chapter 1

Monolithic bearing walls – masonry, stone, concrete

Masonry is the collective term used to describe wall construction using brick, stone, concrete block, or other stacked modular elements. Each component part is set upon another. Using the weight of modular elements to counteract external loads the construction is structurally efficient in compression but poorer at resisting perpendicular tension forces. For the purposes of discussion we will consider bearing wall construction assembled from many small pieces, like brick or stone masonry, to be monolithic, like poured concrete walls, since the structural behavior is essentially similar to that of a single cantilevered plane, conceptually.

Masonry is one of the most ancient and durable construction methods and requires intensive hand labor. Each worker can, however, lay the constituent elements if size and weight permit; therefore it is an economical construction system. Rich and varied effects can be produced from many small pieces, for constructive as well as decorative effect. Stacking of similar elements is the simplest and most ancient of building techniques, and is still widely used around the world. Historical examples and monuments can be found across all cultures with access to stone and brick manufacture. These buildings and structures made of highly durable materials have survived the centuries and show evidence of increasing sophistication and technological innovation in each culture, as the builders explored new methods of achieving structures for cultural, political, and religious glory.

Historical survey

Ancient monuments such as the Great Pyramid at Giza, Egypt, 3000 BC (Figure 1.1), were built with long-lasting masonry to protect the bodies of the ancient rulers after death for return of the soul. These tombs brought glorification to the Egyptian Pharaohs by virtue of the sheer scale and size of the pyramids, made all the more impressive by the limited tools and materials available at the time. There are many theories regarding how the material for the pyramids was transported from faraway quarries and erected. What remains exposed to view today is largely the interior granite structure. Limestone facing panels were originally attached, likely with dowels, to the rough limestone backup blocks largely remaining today, dressed to a smoother, finer finish.

Figure 1.1

Pyramids at Giza

The Great Wall of China, 210 BCE (Figure 1.2), is another example from a different culture where small elements were assembled into a monumentally scaled structure. Rammed earth with masonry rubble infill was used to give internal structure to the fortification (see Figure 1.3). Faced with more durable stone masonry during the Ming dynasty and extended, the wall was used to protect the Chinese against incursions by hostile forces. Besides military control of the border, the wall functioned to provide a continuous path of access for the troops. It also created a physical barrier for trade and an opportunity to control taxation, and became a symbol of unity and vast control, throughout its length.

Figure 1.2

Great Wall of China

The ancient Greeks and Romans used stone and masonry in the construction of monuments, temples, and lesser structures, as well. The permanence of these structures, from the Parthenon from 447 BC through examples of Roman temples and ruins in cities across the Empire surviving today, is a testament to the durability of the simple stacked masonry technique. Roman masonry construction produced excellent examples of brick use at Ostia, a port town near Rome, in the first to second century AD. Brick-faced concrete warehouses, shops, and residences were constructed in this region where the soil was rich in terracotta/brick slay and silt, sand, and gravels (see Figure 1.4). There were no good local marble quarries, only soft tufa and travertine stones, so masonry brick made from locally available materials was assembled using mud between joints as mortar, initially. Limestone was ground into lime powder and used to form an early lime-based mortar. Roman brick was fabricated in squares, cut on both diagonals into right triangles, and then embedded with this improved surface

Figure 1.3

Great Wall of China – restoration view

Figure 1.4

Roman brick – Ostia

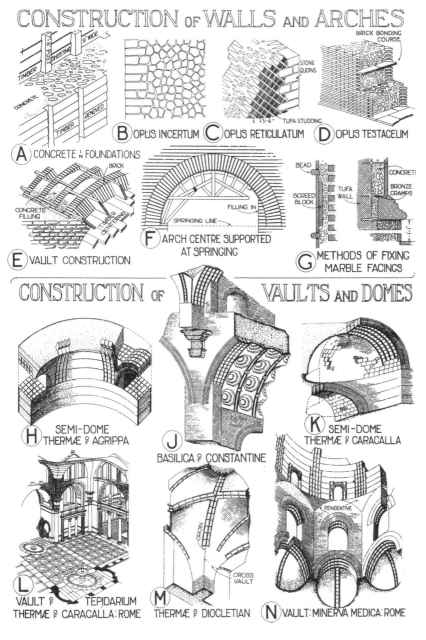

◀ Figure 1.5

Roman brick – construction of walls, arches, vaults, and domes

exposure to rubble fill. As shown in Figure 1.5, this produced a composite brick-faced concrete wall with the more durable brick used only at the surfaces exposed to wear. In some types of brick construction, a leveling course was used to bond the brick and concrete sandwich, a precursor to modern ties. A variety of face-brick insertions into the rubble masonry can be identified, from the more regularized, rectangular shapes of Opus Reticulatum (see Figure 1.6), through rougher irregular-faced masonry, such as Opus Incertum.

▶ Figure 1.6

Opus Reticulatum

The ancient Greeks and Romans had discovered that volcanic pumices and tufa stone, most famously the ones found in Pozzuoli, Naples, and thus pozzollan, could be crushed and ground. When mixed with limestone and burned, the result could be used to produce a harder, stronger, and more adhesive material than lime mortar alone. This material was used as a binder for masonry as well as in the mixture for concrete, or Opus Caementicium (see Figure 1.5). The use of this Roman invention dated to the first century, and it could be poured into forms with permanent masonry components or against removable wood formwork.

The Roman Coliseum, 80 AD, synthesized these structural achievements in one example where brick-faced concrete masonry construction was used to produce multistorey construction with arches and vaulting, buttresses to permit added height, and thickened piers at the base to carry the greatest loads giving expression to structural forces. Figures 1.7 and 1.8 show vaults and arches at the Coliseum. The massive building was used for public displays, lifting the spectators several stories into the air on arched support piers and vaulted roofs, which are now visible in the exposed ruins (Figure 1.9).

The Calidarium at the Forum Baths in Pompeii (Figure 1.10) illustrates another example of Roman construction where concrete-filled brick masonry was used at the piers, walls, and vaults. The rounded Roman arch rotated three hundred sixty degrees became the Roman vault, with an open oculus at the top to allow daylight to penetrate the heavy walls of the volume. The knowledge of concrete construction was lost with the fall of the Roman Empire, and not regained until the end of the eighteenth century. A patent for artificial cement was not established until the early nineteenth century by Joseph Aspdin,[1] who called it "Portland cement" for its resemblance to the British limestone from the region of Portland.

Before this, when control of the Roman Empire shifted east toward Constantinople, Byzantine examples of vaulting spread with the Empire, though adapted to local materials at hand. The dome of the Hagia Sofia, Istanbul,

▲ Figure 1.7

Coliseum, Rome, interior view

◄ Figure 1.8

Coliseum, Rome, view of masonry vault

▲ Figure 1.9

Coliseum, Rome, view of interior

▶ Figure 1.10

Interior view of Calidarium, Forum Baths at Pompeii

537 AD (Figure 1.11), created the largest cathedral in the world for many centuries. Since pozzollan was not available, bricks were used in alternate layers with thick layers of dry mortar, composed of brick fragments and dust, to create the compressive vault geometry. The surfaces of these vaults were covered with stucco and mosaics until earthquakes in 553 and 557 caused damage to the main dome, which collapsed in 558 and required complete reconstruction in the following years before becoming a mosque in the 1450s.

Examples of load-bearing masonry walls abound in the Islamic world. In Bukhara, Uzbekistan, the Tomb of the Samanids, c.900 AD, used both standard and T-shaped bricks (see Figure 1.12). The cube form of the structure is topped by a dome, just as the heavens top the earth, where the cube shape reflects the form of the Kaaba at Mecca. Together, heaven and earth represent the universe. The detail views in Figures 1.13 and 1.14 show how the patterning of the face brick was set to create complex patterns. At the exterior these masonry patterns exploited the changing angles of sunlight to create an active surface texture using light and shadow.

The Great Mosque of Isfahan, 1611–1630 (Figure 1.15), another example of load-bearing masonry construction, incorporated a 52m outer shell with

◀ Figure 1.11

Hagia Sofia, interior view of dome

◀ Figure 1.12

Tomb of the Samanids, exterior view

▲ Figure 1.13

Tomb of the Samanids, detail inside dome

▼ Figure 1.14

Tomb of the Samanids, detail of windows and other exterior brickwork

◀ Figure 1.15

Great Mosque, Isfahan, detail of dome

non-structural vaults, squinch, and decorative tile over brick. Elaborate geometric decoration can also be seen on the muqarnas vaults. These increasingly large, daringly long spans using domed masonry construction sheltered the faithful during religious gatherings, and made use of the muqarnas vaults with elaborate tilework decoration using complex geometries. Figure 1.16 shows an exterior view of tilework. Used as academies of learning, as well as religious instruction, the structures and their surfaces reflect the sophistication of the mathematical, technical, and philosophical studies undertaken inside.

In medieval Europe, where verticality came to be fashionable, developments from the ancient Roman rounded arch still used throughout the early Middle Ages led to the pointed Gothic arch. Pointed spires were set on side wall reinforcement buttresses at the sides of the main arch, expressing a need to counterweight lateral wind forces by using the downward gravity action of the masonry spires. Heavy foundations were required below ground to support these structures (see Figure 1.17). At the Cathedral of Notre Dame, Paris, 1250, flying buttresses were used to further relieve the vertical wall from load-bearing functions, bracing the vertical walls against the outward thrust of the vault (see Figures 1.18 and 1.19). Views from the exterior at the curved choir, behind the altar, show the flying buttresses, secondary anchors against the lateral thrust of the taller central spans.

Robert Mark's structural modeling using scaled models of cathedrals and other structures, cut from fresh sheets of a transparent material, drilled and loaded with weights (upside down to simulate gravity loads, with added weights pulling sideways to simulate lateral forces and wind pressures), were then viewed through polarized lenses to produce a visual expression of the stresses acting on a section of a structure, where color shifts could be counted to extrapolate stresses. Figure 1.20 shows an example of the photoelastic model study for the cathedral at Beauvais, France. The high Gothic choir, completed in 1272, was being extended into the nave when, in 1284, a collapse occurred. The overstressed masonry is shown under the influence of lateral wind load in the structural study for the largest, and perhaps most daring, Gothic vault in the world.

▶ Figure 1.16

Great Mosque, Isfahan, detail of courtyard

▲ Figure 1.17

Thick masonry wall – Crypt at Les Invalides

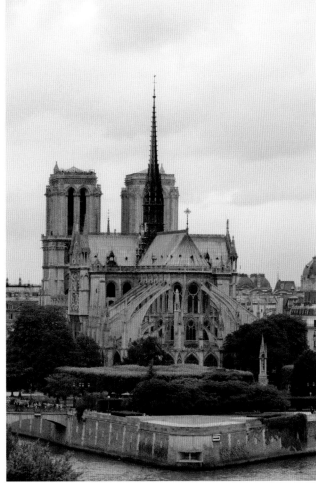

▲ Figure 1.18

Notre Dame de Paris

In the Americas, Machu Picchu provides an example of a pre-Columbian Inca site in the Cusco region of what is today Peru. Constructed in the fifteenth century, the city was built using various types of coarse ashlar-cut stone through rough-dressed rubble stone on a steeply sloped site, shown in Figure 1.21. All of the masonry was dry set, erected without mortar, using carefully profiled stone (see Figure 1.22), worked to achieve full load transfer of the weight of the buildings without relying on adhesion of joint filler material, incorporating openings limited to relatively small trapezoidal spans. The masonry was used for retaining walls as well, terracing the fertile soil of the city for crops as well as defense against any possible intruders.

The cliff dwellings of the ancestral Pueblans, or Anasazi, at Mesa Verde in Colorado, USA, likewise used the natural topography for defense and in this

▶ Figure 1.19

Notre Dame de Paris

▶ Figure 1.20

Photoelastic interference pattern produced by simulated wind loading in a model of the Beauvais Cathedral choir structure

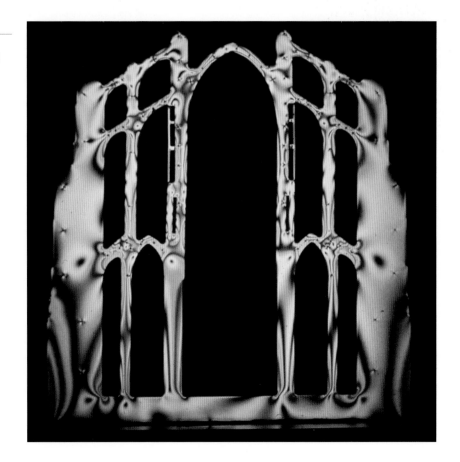

CHAPTER 1 Monolithic bearing walls – masonry, stone, concrete

▲ Figure 1.21

Machu Picchu, Cusco region, Peru

◀ Figure 1.22

Machu Picchu, Cusco region, Peru

▶ Figure 1.23

Mesa Verde, Cliff Palace, general view

▶ Figure 1.24

Mesa Verde, Cliff Palace

case, shelter. Constructed between 600 and 1300 AD, the natural caves and stone ledges provided shelter for the generations who farmed the "mesa," or table-top lands, hunted, and produced both basket weavings and clay pottery. The Cliff Palace (Figures 1.23 and 1.24), a multistorey complex of houses, ritual spaces, and storage rooms, were built into the sandstone ledges of the canyon walls, protected by the rock overhangs. Facing south, to capture the heat of the low daytime sun in winter, the sandstone walls were built into the natural shallow

◀ Figure 1.25

Mesa Verda, Cliff Palace, detail of restored masonry

caves with adobe (clay, sand, water, and organic reinforcing straw). Figure 1.25 shows a detail view of a reconstructed wall. With sufficient overhang, the high summer sun would throw the dwellings into shade, but allow breezes from the valley to cool the inhabitants.

* * *

Masonry was used for millennia to create permanent structures of all kinds. Historically, brick and stone comprised the most costly construction type, and therefore was selected for religious and cultural buildings of greatest importance. Stacking smaller elements (or pouring concrete) to provide support for upper floors and to resist wind forces against the side walls was the only system of construction until the modern era. Limits to the gravity-loaded construction were explored in the Monadnock Building, 1891, Burnham and Root, constructed after the Great Fire of Chicago in 1871 (Figure 1.26). In response to the desire to construct fire-resistant buildings, especially for offices, warehouses, and printing presses where large quantities of paper and other combustible materials were inevitably kept, brick and terracotta construction became common, even with the early cast-iron buildings foreshadowing the future. The Monadnock's building section (Figures 1.27 and 1.28) shows how massive the unreinforced wall had to become, in order to carry the loads associated with a tall building – 18" thick at the top, but 6' thick at the base. Rising 16 stories, it reached the limit of unreinforced masonry bearing wall design.

▲ Figure 1.26

Monadnock Building, exterior view

◀ Figure 1.27

Monadnock Building, view of thickened masonry at base

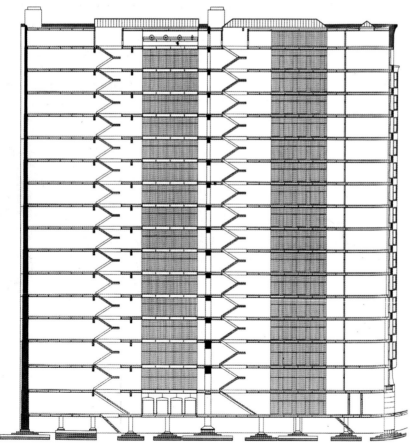

◀ Figure 1.28

Monadnock Building, building section

Structural requirements of load-bearing walls

As with the Monadnock building, thickness of the load-bearing wall is dictated by the unbraced vertical length, solidity, and reinforcing of the wall. Rule of thumb:

- Roughly 12' thick wall might be sufficient for 35' in height, then 4" for each additional 35'
- Solid wall height = 20 × thickness (T)
- Hollow wall height = 18 × T
- Non-load-bearing wall height = 36 × T

Exceptions to the rule of thumb for stacked, simple unreinforced walls include using geometry to increase the effective strength of the construction. Like coiling a sheet of paper into a cylinder, which allows it to support much greater axial load (it can support a book on top of the cylinder), folding or curving the surface can add to the material capacity to resist loads. Architectural examples of this geometric trick include Thomas Jefferson's serpentine garden walls at the University of Virginia, where a curvature in the shape of the plane makes a wall of single-brick (or wythe) thickness stable against overturning. In his sketch studies (Figure 1.29), we see him considering the reinforcing effect of incorporating piers with deeper profiles to stabilize the web of wall in between such lateral reinforcing locations. Alternatively, Jefferson drew simple curved shapes that would stiffen a single wythe of brick masonry that could resist overturning lateral forces by virtue of the form. At the University of Virginia, the dividing walls between gardens (Figure 1.30) are slender but stable, thanks to the sine curve plan profile of the conventional bricks.

Eladio Dieste's projects in Uruguay such as the Church of Christ the Worker in Atlántida, 1961, exploited the same principles as the serpentine walls at the University of Virginia on a larger scale (see Figures 1.31, 1.32, and 1.33). The steel-reinforced undulating walls were constructed three wythes thick, as can be seen in the construction photos shown as Figures 1.34 and 1.35. The compound curved surface rises from a simple line, at the foundation, into ruffled profiles at

▼ Figure 1.29

Thomas Jefferson, sketch of brick wall with piers or serpentine to provide added support and allow greater height

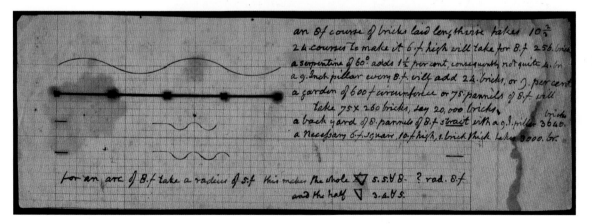

▲ Figure 1.30

Thomas Jefferson, University of Virginia, serpentine garden walls

◀ Figure 1.31

Church of Christ the Worker, Atlántida, Uruguay – side elevation

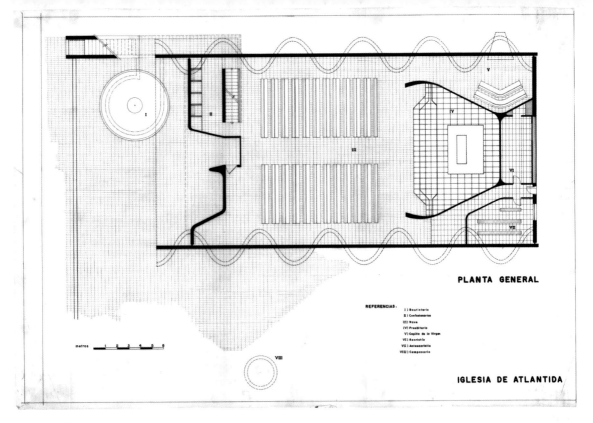

▲ Figure 1.32
Church of Christ the Worker, Atlántida, Uruguay, Eladio Dieste – plan

▼ Figure 1.33
Church of Christ the Worker, Atlántida, Uruguay – section

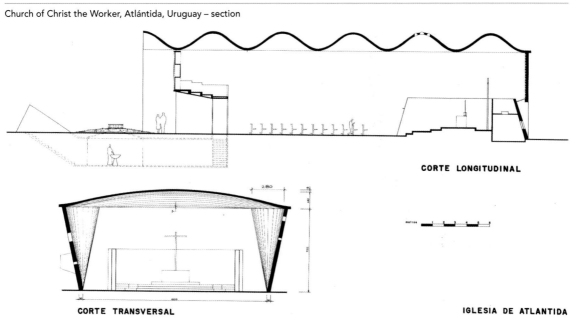

CHAPTER 1 Monolithic bearing walls – masonry, stone, concrete

◀ Figure 1.34

Church of Christ the Worker, Atlántida, Uruguay – construction view with vertical reinforcing bars

▶ Figure 1.35

Church of Christ the Worker, Atlántida, Uruguay – construction of three-wythe-thick monolithic wall

◀ Figure 1.36

Church of Christ the Worker, Atlántida, Uruguay – interior detail view

the roof eaves on both long sides of the church nave. The curved roof surface meets the double curved side walls very carefully (see Figure 1.36). Throughout the complex, simple brick modules are used in conventional and unusual ways to play with light and shadow (see Figure 1.37). At the bell tower, curvature is created with regular, rectilinear brick using open alternating modules to screen the interior from sunlight but let the sound of the bells out.

As the previous examples have shown, the history of architecture is rooted in the monolithic bearing wall construction type. Modern day uses of this wall type abound as well, though limitations of heights and spans for braced and unbraced construction dictate the scale and height for economically effective load-bearing monolithic wall buildings used today. Recent brick and concrete

▲ Figure 1.37

Church of Christ the Worker, Atlántida, Uruguay – campanile interior

masonry examples include infill structures, where bearing walls establish a module of planes with infill facade closure at each end, to define a party wall multi-family house, hotel, or other party wall system. Industrial/commercial examples include big box stores, for example, in some parts of North America.

A large number of local industries are required in order to manufacture brick, though each could be accomplished with a single worker, historically. Raw materials consist of clay and silt, which must be dug up from the ground, crushed, ground smooth, screened to a fine consistency, then mixed with a controlled quantity of water and either pressed into a mold (historic) or extruded (modern industrial).

The appearance, strength, and performance of the built structure are all affected by the manufacture of the component units: handmade versus machine-made; textures achieved by water content. The brick is dried in a low-temperature oven, or kiln, before being fired at high temperature, then cooled slowly for between 40 and 150 hours. Substantial shrinkage occurs, and with large production runs these are calibrated to achieve the desired final sizes (see Figure 1.38).

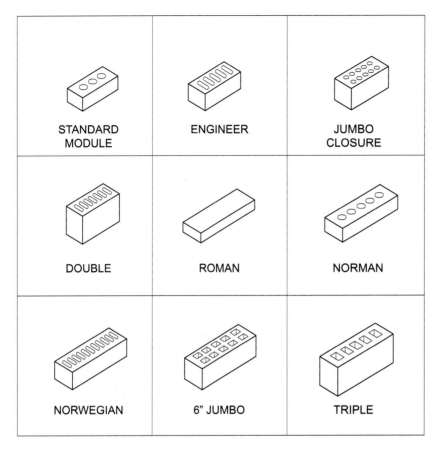

◀ Figure 1.38

Brick types

Color can vary based on the chemical composition of the clay substrate as well as the chemistry of the kiln-firing process. Smaller bricks are less likely to crack in firing. Higher temperatures can produce darker colors. Since the brick is produced by fire, it is naturally somewhat fire-resistant. Bricks can be shaped by hand, molded by pressing into forms, or extruding and cutting of desired lengths. Modular brick dimensions are designed to meet even coursing dimensions when combined with mortar joint thicknesses of ¼"–½", as a rule. Most bricks manufactured in North America (Standard brick, Modular brick, and Norman brick) measure 8" over three courses high, including mortar. Special ornamental types are made to different dimensions, though the conveniences of modularized accessories and backup materials require coordination.

Of the many sizes and shapes available, Modular Brick, the most common, has the following properties:

- 4" × 8" nominal outside dimensions after firing
- 3 courses high = 8"
- Actual size is ⅜" less (to accommodate the typical mortar joint dimension, just under ⅜")
- 2¼"H × 3⅝"W × 7⅝"L, approximately.

▶ Figure 1.39

Robie House – exterior view

When considering brick size in design, it may be worth remembering that jumbo brick can be more economical, and is therefore often considered for developer projects which are highly price driven. Availability of specialty brick sizes, especially for smaller projects, may negate possible savings in labor and mortar hand-work, however.

Grades and types of brick

There are three grades of brick based on resistance to weathering, and three types of facing bricks, based on degree of uniformity: shape, size, texture, color. The standards for these types of brick are established by the American Society for Testing and Materials. Industry associations and sales representatives; the Brick Institute of America; online construction indexes; and brick manufacturers all offer extensive support and technical information.

An example of special brick shapes and mortar use in a historical example is illustrated by Frank Lloyd Wright's Robie House, Chicago, IL (Figure 1.39). The use of Roman brick in this case accentuated the horizontality of the building

▲ Figure 1.40

Robie House – detail showing Roman brick with colored masonry to emphasize horizontality

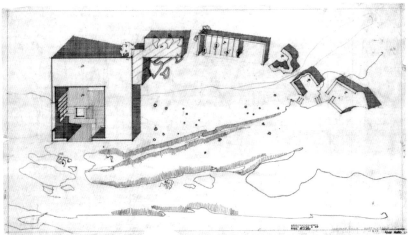

◀ Figure 1.41

Alvar Aalto, Experimental House, plan

profile, low slung and paralleling the ground plane, like the Midwestern prairie, thus suitable for the "Prairie Style" of FLW's imagination. The horizontally laid elongated Roman brick was further exaggerated by the deeper rake of the mortar at the horizontal joints, against the flush vertical joints, which were also a different tint (see Figure 1.40). This level of craftsmanship on the part of the mason is an impressive testament to both the owner's commitment to the architect, which may not be easily matched today, and that architect's powers of persuasion.

Alvar Aalto's Experimental House in Muuratsalo, Finland, 1953, near Jyväsklä in the islands around Säynätsalo, was his own weekend house. A series of small outbuildings fanned around the view, clustered in a little community of cabins facing south (see Figure 1.41 for plan). Constructed with monolithic

▲ Figure 1.42

Alvar Aalto, Experimental House, exterior view

open courtyard walls surrounding a firepit (Figure 1.42), with views through the gap in the wall towards the water, the outside face of the courtyard wall were painted white. The interior was used to study light and shadow in the textures of various brick bonds and patterns, in mock-up panels of different brick pointing, brick sizes, ceramics, and finishing techniques that Aalto could observe over the course of the seasons and under changing light conditions (see Figure 1.43). This is undoubtedly the best method for selection of brick, configuration of special coursing, and observation of mortar color against brick color in wet and dry conditions. Using corners of the courtyard and house, viewable from even the master bedroom, at all times of day and night, would have given the architect the most information on the effectiveness of each experiment in material and assembly for the brick masonry.

Concrete masonry units (CMU)

Concrete masonry units, commonly called "Concrete Block", are fabricated to 4", 6", and 12" widths, though the most common overall nominal size is 8" × 8" × 16" (actual dimensions again are ⅜" less on each side to account for mortar joints). Unlike brick, which is laid with one hand, concrete masonry units, or "CMU", require two hands to lift and set. There is a modular relationship between the most commonly used brick and block dimensions: one block is three bricks in height. Manufactured using a stiff concrete mix vibrated into molds, turned out onto racks, and steamed, the concrete cures by hydration.

◀ Figure 1.43

Alvar Aalto, Experimental House, interior view of brick experiments at courtyard wall

▶ Figure 1.44

Louis Kahn, Trenton Bath House

▶ Figure 1.45

Louis Kahn, Trenton Bath House

The CMU, being reinforced with steel horizontal ladder trusses (commonly every third course of CMU) with grouted cores and steel vertical reinforcing, offers significant structural performance and is often used as the backup wythe to brick-faced veneer wall construction.

Louis Kahn's Jewish Community Center Bath House, Trenton, NJ, 1955, created shelter and enclosure for the swimming pool's changing rooms using CMUs (see Figure 1.44). The open-air cluster of huts suggest a monumental ruin around a symbolic central hearth (Figure 1.45) was built entirely of unfinished single-wythe CMU. In principle, CMU is the more economical construction type since it is quicker to erect. Block can be produced to a given strength requirement by being dried to specific water content. Hollow cores of the units allow for insertion of reinforcing steel and grout, for improved strength and performance under tension. While blocks can be pigmented and textured, or faced with either cement plaster or exterior insulating finish systems, the Bath House is formed from a series of simple block enclosures, open at the eaves, with hipped roof structures sitting on the corner entry elements. The complex was recently purchased by the local preservation trust and government, circumventing previous demolition plans.

Mortar

Mortar does not "cement" the wall together, but rather spreads load, seals joints, and facilitates bearing on irregular surfaces, acting to cushion the masonry. Mortar is mixed from a combination of Portland cement, hydrated lime, inert aggregates like sand, and water. The Portland cement acts as the bonding agent. In early examples, where just lime was used, slow and uneven curing was inevitable, but the lime reduces strength while improving cure. Four typical strengths can be specified today, depending on project requirements. After mixing, mortar must be constantly worked and can only be used for 90 minutes unless it is rehydrated, thus a two-person team is often the minimum, where the mason setting brick is kept productive by virtue of an assistant tending the mortar. Since 20 percent of a wall area is likely to be mortar, color selection also becomes significant – under wet and dry conditions if color match is important.

Brick terminology

The following key terms are widely used in brick naming conventions (see Figure 1.46):

- Course/Wythe: One horizontal row of brick construction/vertical column
- Stretcher: Brick laid with length parallel to the face of the wall, lying down
- Header: Brick laid with end parallel to face, bonding two wythes together
- Soldier: Brick set on end, standing up straight facing outward "like a Soldier" (for window heads or arches, for emphasis)
- Rowlock: Brick laid on its face, ends parallel (for wall caps)
- Bed Joint: Joint between courses of brick
- Head Joint: Joint between bricks
- Collar Joint: Joint between courses.

Position of Brick in the Wall
Interesting brickwork design effects can be achieved by placing brick in other than the normal stretcher position, as illustrated below. Regardless of how the brick is installed in the wall, size dimensions (thickness, height and length of the unit) are always assigned from the stretched position.

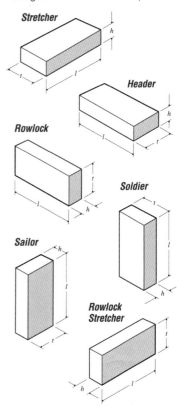

Glen-Gery Brick Sizes for Standard Shapes

	t	h	l
STANDARD	3-5/8"	2-1/4"	8"
STANDARD MOD	3-5/8"	2-1/4"	7-5/8"
ENGINEER STD	3-5/8"	2-3/4"	8"
ENGINEER MOD	3-5/8"	2-3/4"	7-5/8"
ECONOMO	3-5/8"	3-5/8"	7-5/8"
UTILITY	3-5/8"	3-5/8"	11-5/8"
*HMOS	4"	2-3/4"	8-1/2"

*Handmade Oversize

▲ Figure 1.46

Brick shapes and terminology

▶ Figure 1.47

Brick coursing bonds

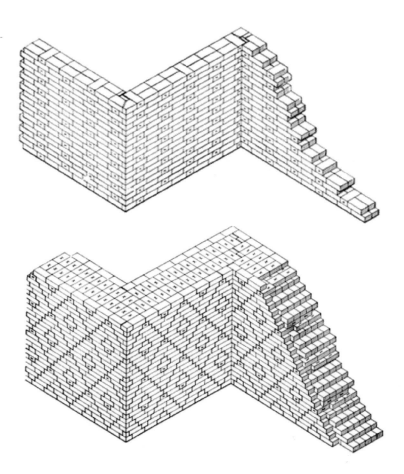

Structural bonds

While there are endless permutations of brick combinations and patterns, the aesthetic impact of tying bricks together using stacking patterns is called a "bond", the most common of which bear highlighting (see Figure 1.47):

- Running bond: All stretchers, with staggered joints (typically one wythe thick, with no bonding course perpendicular to the face)
- Common bond: Running bond with a header course every sixth course (likely spanning a two-wythe-thick wall)
- English bond: Alternating course of headers and stretchers
- Flemish bond: Alternating headers and stretchers within each course.

Bearing wall: masonry

If we look at a representative modern building wall section through a monolithic masonry wall (Figure 1.48), we see two wythes, set perhaps 2"–4" apart, with

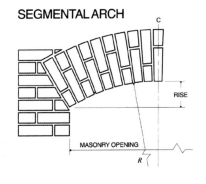
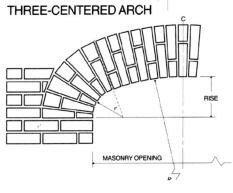
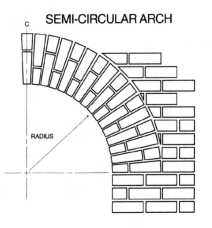
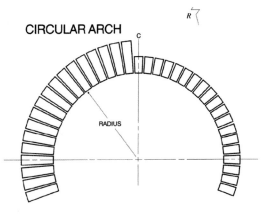
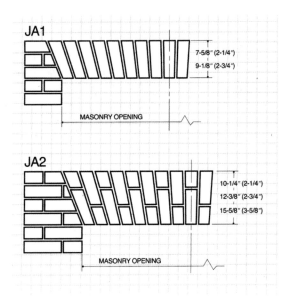
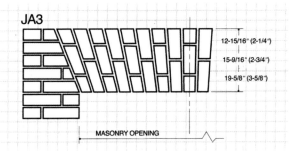

Figure 1.48

Wall section – monolithic brick bearing wall construction

steel reinforcing bars ("rebar") in the cavity. Galvanized steel wire ties spanning across would be used to tie the two wythes together and resist any outward pressures from grout, filling the cavity like a header brick but better at resisting tensile forces. Grout, consisting of Portland cement, sand, and water, could be poured between the wythes to provide stability for the reinforcing bars, with a surprisingly thin final assembly capable of providing bearing for floor and roof construction.

Masonry construction technique

Laying of concrete block and brick walls begins at the corners, to ensure that the finished construction is plumb, level, and true. Corners are set with lead lines, guideline strings established by the most experienced mason on the job, so that the infill between corners and between guidelines can be executed by less experienced masons. Brick is set with one hand, if standard size, unlike CMU, which requires two hands to lift and set. Edges of each brick are buttered with mortar, and the wall is built up to the lead lines. Residual mortar must be removed immediately to prevent curing in place and staining the finished face of the work. Unfinished work should be covered at the end of every work session or workday to prevent water from collecting within the unfinished work.

Where care is not taken to prevent water intrusion efflorescence can occur, especially over time. White soluble salt deposits leach out of the brick by water penetrating the surface and pulling the soluble material out of the brick. To minimize the chances of efflorescence, impede water intrusion: fill joints solidly, slope wall coping toward the roof and away from the vertical face of the exterior wall, use overhangs, and protect the top of the wall during construction.

Dimensioning

When designing for masonry construction, it becomes important to understand the module and take into account the size of the actual masonry units selected. Dimensions of walls should ideally be based on full units, with the dimension of the bricks plus one joint for an overall clean dimension. Likewise, openings for windows and doors, or "masonry openings", should be sized to a full brick dimension plus one joint.

Jointing

It is important to ensure that the mortar has contacted both sides of the joint it fills before it sets. Tooling, using a tool to push the mortar into the gap, can be done with a concave or V profile, in section. These are likely ideal since they naturally shed water, which helps avoid capillary action in any cracks, thus best resisting freeze-thaw cycles.

Reinforcement

Steel, being good in tension, can be used to supplement the masonry, which is good in compression, bad in tension. Rebar can be placed in the brick collar joint, horizontally, and in CMU cores, vertically. In a brick-faced-block wall, horizontal wire ties spanning across both wythes can be placed every 16" horizontally, engaging both standard brick and CMU modules. Galvanized ladder trusses grouted within the bed joint every three courses of block would then be every 24" vertically, or eight brick courses. Reinforcing keeps the tie adjacent to blocks and keeps vertical blocks stacked, preventing them from tipping over.

Openings

To span over openings in a brick wall, lintels can be used to provide a support for upper masonry and transfer loads to either side, given sufficient bearing surface at the sides. Piers of brick support the spanning member, which can be stone (can be decorative as well as functional, but stone can be expensive and weak in tension), precast concrete, steel angle, beam, or some combination of concealed or exposed composite, reinforced, grouted "bond beam". Alternatively, arches can be used to create a lintel out of smaller masonry elements (see Figure 1.49). Bricks can be formed or cut, called "gauged," to a wedge shape, and/or mortar can be used to fill the spaces between rectangular units. Curved or pointed or flat arches, of course, support the construction above through a careful balancing game, without additional reinforcement.

Expansion/contraction

To address the expansion and contraction of the masonry, joints with a soft, compressible and expandable filler are installed at offsets and junctions of walls, for instance in buildings with an L-, T-, or U-shaped floor plan. These joints must provide a complete separation of all wythes in a wall assembly, with no bonding. Filler material has evolved from cork, cocoa matting, etc., to modern-day sealants. See Chapter 4 for detailed discussion.

Insulation

To achieve a habitable enclosure, the modern exterior building skin should provide insulation against outside temperatures. If the outdoor climate is cold, the exterior enclosure needs to retain heat within. The architecture of a given region typically is geared to address the predominant source of discomfort across the annual cycle, so for a cold, Northern climate, sufficient insulation and heat retention, combined with a satisfactory level of fresh air heated to a comfortable temperature, becomes the design goal. For a hot climate, sufficient shade to prevent the interior from growing too hot and sufficient insulation to

▶ Figure 1.49

Flat jack arches and rounded arches in brick

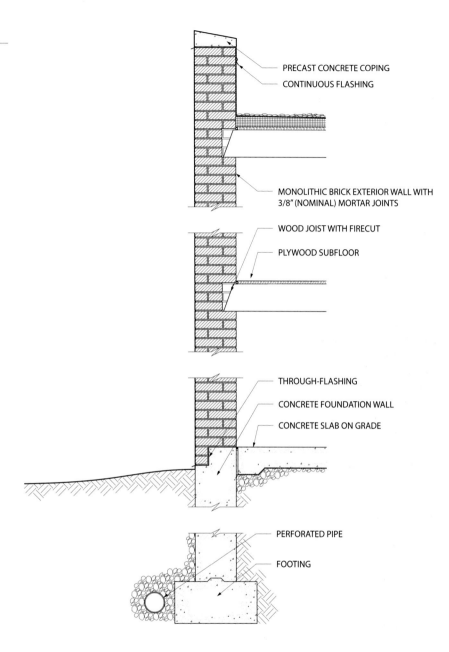

retain cool air created by mechanical means, within, could be the predominant design scenario. While extreme climates can be made habitable – at the Antarctic or for the millions of inhabitants of equatorial regions – many design conditions incorporate some of both heating and cooling conditions, so the exterior building envelope must address both in the thermal design of the assembly.

In addition to insulating properties of wall assemblies, we also consider humidity and vapor movement. A vapor barrier installed on the warm side of insulation can prevent migration of moist air to the interior of a building, though condensation forming inside insulation where it is not permitted to dry out can damage the insulation, much less invite growth of mold. Monolithic masonry is difficult to insulate within the construction, and historically has no vapor barrier at all. This is especially a problem in high-humidity climates or conditions where building use causes higher levels of interior humidity. Insulation at the interior face of the wall can be difficult to install in such a way as to maintain continuity of both insulation and vapor barrier, since the floor slabs break the line of interior insulation (see Figure 1.49), though in a hot or tropical climate where the vapor barrier is desirable at the exterior face it may be easier to achieve continuity.

Flashing

Flashing is a waterproof, formable material such as metal (stainless steel, copper, zinc, etc.); waterproofed woven fiber, felt, plastic; or man-made rubber. In monolithic walls, flashing installed in stepped configuration through brick courses can drain moisture to the exterior if small holes are installed in the outer wythe to permit moisture to drip out through weep holes. There are many options and products to further improve weep holes: plastic tubes can be used to set a minimum width, or sometimes can be a full brick high, a long oval, set into the head joint. If too large, pests and insects are liable to obstruct the opening, so the manufactured weep tubes are sometimes given a filler material. Cotton rope was used historically to conduct moisture out, like a wick, and can also be offered as weep filler.

Flashing should be installed above and below every opening. Jamb flashing at either side must be overlapped so that moisture draining down falls onto the sill flashing and is conducted outward. Construction therefore is likely to work from bottom up, like the masonry wall itself. Windows are inset within these flashed openings, if not already manufactured with integral flashing components which should be installed with overlaps and tucked into adjacent membranes to ensure continuity of vapor barrier and water flow down- and outwards. A wealth of educational opportunities exists in failure investigations, whether due to missing flashing or improperly oriented materials introducing unnecessary water into the wall.

Bearing walls: stone and concrete

Like masonry units, stone and concrete bearing wall construction can be used individually or in combination to support floors and roofs as well as provide the exterior enclosure of a building. In the example wall section shown in Figure 1.50, a rough stone wall shows the top of a stepped concrete foundation (stepped to encourage positive drainage of moisture within the wall out and away from the interior) supporting a wall assembled from blocks of stone (rough stones stacked to achieve a relatively uniform surface exposed to view, both inside and out). The slab-on-grade concrete floor is, in this case, not tied into the wall structure,

▶ Figure 1.50

Wall section – monolithic stone bearing wall construction

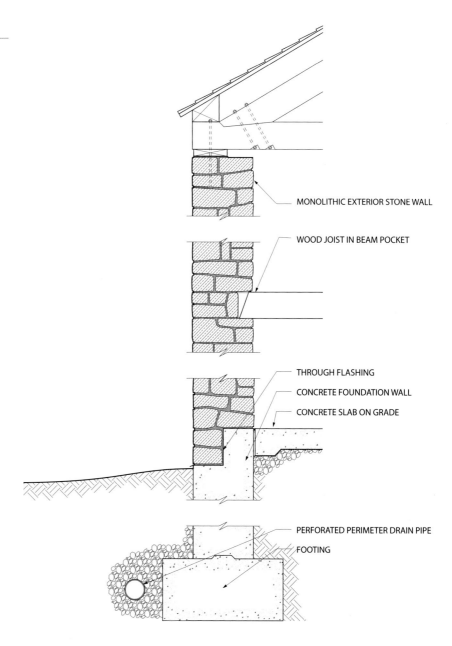

though it could be. The wall section illustrates an example where the next floor above is supported on the stone wall, like the roof. In this example section, the wall construction is simultaneously acting as both the shield for wind, rain, and the elements, as well as providing the structural support for the building above. This is the historic, traditional role of an exterior wall, acting as load-bearing element as well as environmental buffer.

The type of stone used for load-bearing construction can be chosen for both visual as well as structural reasons. Knowing what the material can bear, in terms of strength and how it will behave under different load conditions, is dependent on a number of factors: type of stone, consistency from the quarry, and how it is cut. Being a natural material quarried from the ground, the method in which the stone was formed affects its strength and performance under load. Stone can be divided into three types – igneous, sedimentary, and metamorphic. Igneous rock is stone that was deposited in a molten state, like granites, and is therefore very uniform and monolithic relative to most other stone. Sedimentary stone is multiple layers of material, deposited by the action of gravity, wind, or water, for example limestone or sandstone, and can react quite differently if used perpendicularly to the direction of the layering, rather than parallel to the rift/layering. Metamorphic rock could be either of the previous stone types that was subsequently transformed by heat or pressure into a different material, for example slate or marble.

In all cases, stone properties to consider include: permeability (slate being one of the least permeable, sandstone one of the most); porosity (the absorption of marble, for instance, can be twice that of granite, sandstone can be one of the most absorptive); density/weight (all types can be somewhat similar and rather heavy); and shear strength (granite and marble can be fairly similar, though every stone is likely to behave quite differently, being created from different processes). Figure 1.51 shows different stone types and properties. To estimate the likelihood of consistency within one batch, for use on one large building project, for example, it is possible to test for these properties and establish the statistical likelihood of variation.

In addition to these factors, the way a stone is fabricated can also affect the performance when used over the life of a building. Maximum sizes may be limited by the strength of the stone as well as dimensions that can be reasonably taken out of the ground and transported. Minimum thicknesses may be dictated by strength. Weaker limestones may be difficult to manipulate if too thin, stronger granites can be slabbed into much thinner dimensions if service loads permit. While in general it could be expected that stronger stones, like granite, could be thinner, and weaker stones, like limestone, would have to be thicker, all dimensions are subject to variation based on the type of stone, loads to which it will be subject, and method by which it could be supported or anchored.

Stone can be fabricated into different shapes, from very irregular profiles with little applied treatment through elaborately worked profiles and forms. As quarried, blocks of stone can be quite large. Subsequent cutting and fabrication brings the natural material into shapes usable for construction purposes, so the initial block dimensions set maximum architectural use, with some waste at the weakened edges, especially for softer stone. Figure 1.52 shows sandstone blocks from an abandoned open quarry. Cut blocks can be shaped into large and small stones. Stone can be cut into irregular shapes or smoother, rectilinear profiles. All fabricated stone can be dressed with a variety of surface treatments in a range of rough, nearly untreated for a heavier appearance, through more geometrically surfaced hammered finishes, all the way to honed or even polished finish treatments, provided the stone can withstand the working without losing strength or excess thickness. Slabbed stone, especially the more durable granites, can be passed under open flame, which causes the top surface to

▶ Figure 1.51

Stone types and properties

PHYSICAL PROPERTIES OF COMMONLY USED BUILDING STONES					
PHYSICAL PROPERTY	GRANITE	LIMESTONE	MARBLE	SANDSTONE	SLATE
COMPRESSIVE STRENGTH, MIN., PSI	19,000	1,800 (A) 4,000 (B) 8,000 (C)	7,500	2,000 (D) 10,000 (E) 20,000 (F)	NOT LISTED
MODULES OF RUPTURE, MIN., PSI	1,500	400 (A) 500 (B) 1,000 (C)	1,000	300 (D) 1,000 (E) 2,000 (F)	9,000 (G) 7,200 (H)
ABSORPTION BY WEIGHT, MAX., PERCENT	0.40	12.0 (A) 7.5 (B) 3.0 (C)	0.75	20 (D) 3 (E) 1 (F)	0.25
DENSITY, MIN., PEF	160	110 (A) 135 (B) 160 (C)	162 (J) 175 (K) 168 (L) 144 (M)	140 (D) 150 (E) 160 (F)	NOT LISTED
ABRASION RESISTANCE MIN., HARDNESS	NOT YET ESTABLISHED	10	10	8	8
COEFFICIENT OF THERMAL EXPANSION IN. PER. DEGREE F	0.0000063 TO 0.0000090	0.0000024 TO 0.0000030	0.00000369 TO 0.00001230	0.0000050 TO 0.00000120	0.0000094 TO 0.00000120
MODULUS OF ELASTICITY, PSI	5,700,000 TO 8,200,000	3,300,000 TO 5,400,000	1,970,000 TO 14,850,000	1,900,000 TO 7,700,00,000	9,800,000 TO 18,000,000
ULTIMATE SHEAR STRENGTH, PSI	2,000 TO 4,800	900 TO 1,800	1,638 TO 4,812	300 TO 3,000	2,000 TO 3,600
ULTIMATE TENSILE STRENGTH, PSI	600 TO 1,000	300 TO 715	150 TO 2,300	280 TO 500	3,000 TO 4,300
AVAILABLE FINISHES	POLISHED, HONED, RUBBED, SHOT GROUND, THERMAL, SAND BLASTED, CUT, SAWN	SMOOTH, PLUCKED, MACHINE TOOLED, CHAT SAWED, SHOT SAWED, SPLIT FACE, ROCK FACE, CUSTOM	POLISHED, HONED, SANDBLASTED, ABRASIVE	SPLIT FACE, RUSTIC FACE, PITCHED FACE, CHAT SAWED, SAND, SAWED	NATURAL, CLEFT, SAND RUBBED, HONED
STANDARD THICKNESS FOR VENEERS, IN.	1 1/4	NONE	7/8, 1 1/4, 1 1/2, 2	NONE	1, 1 1/4, 1 1/2

A. LOW DENSITY, 110 PEF TO 135 PEF
B. MEDIUM DENSITY, GREATER THAN 135 PEF AND NOT GREATER THAN 160 PEF
C. HIGH DENSITY, GREATER THAN 160 PEF
D. SANDSTONE
E. QUARTZITE SANDSTONE
F. QUARTZITE
G. ACROSS GRAIN
H. ALONG GRAIN
I. FOR SLATE ONLY
J. CALCITE MARBLE
K. DOLOMITE MARBLE
L. SERPENTINE MARBLE
M. TRAVERTINE MARBLE
N. PERTAINS ONLY TO STONES SUBJECTED TO FOOT TRAFFIC

PERMEABILITY OF COMMERICAL BUILDING STONES (CU IN. PER SQ FT PER HR FOR 1/2 THICKNESS)			
KIND OF STONE	PRESSURE, PSI		
	1.2	50	100
GRANITE	0.6–0.8	0.11	0.28
LIMESTONE	0.36–2.24	4.2–44.80	0.9–109
MARBLE	0.06–0.35	1.3–16.8	0.9–28.0
SANDSTONE	4.2–174.0	51.2	221
SLATE	0.006–0.008	0.09–0.11	0.11

◀ Figure 1.52

Sandstone quarry, Red Rock Canyon, 1880s

◀ Figure 1.53

Ames Free Library, North Easton, MA, H.H. Richardson, exterior view

break off in an irregular pattern. Flamed finishes expose the interior of the stone to view, and like many surface treatments can significantly change color perception of the substrate. Fabricated stone can be assembled in any number of coursing methods, from the more rustic rubble or rough ashlar coursing, as used in many Richardsonian Romanesque buildings of the late 1800s, through finer and more regular coursings.

Examples of how stone was used in North America include Ames Free Library, North Easton, Massachusetts, an H.H. Richardson building where stone was used to emphasize the sense of stability and mass (see Figure 1.53). Tapered, or "battered", walls also incorporated graduated colors and textures from rough at the bottom to smooth at the top of the building to seat the building in its landscape. The windows are inset deep into the walls, concentrated at the top of the wall to provide glare-free clerestory daylighting. Wall sections taken from Ed Ford's *Details of Modern Architecture* (1990) show that the load-bearing walls were faced with rough ashlar Longmeadow stone with backup brick (to minimize cost). Figure 1.54 shows an axonometric view illustrating these

▶ Figure 1.54

Oliver Ames Free Library, 53 Main Street, North Easton, Bristol County, MA, axonometric showing wall assembly. A – Roof construction: flat tiles nailed to wood boards spanning between rafters. B – 8" rafters spaced at 1'8" on center and supported by masonry wall. C – Wood sill. This provides a level nailing surface for the rafters and transfers their load to the masonry wall. D – Ceiling of 1 × 3 tongue-and-groove boards. E – Gray Milford granite and pink granite wall. F – Air space to protect books from moisture. G – Longmeadow sandstone trim. The carved and dressed portions of the wall are of sandstone, which is softer and easier to carve than the granite. H – Brick. As much of the wall as possible is made of common brick, laid simultaneously and interlocking with the stone (brick being much less expensive). I – Floor construction: 2 × 12 wood joists 12" on center supporting wood finish floor. J – Foundation. The wall is thickened at the base and made of the same stone as the wall although it is dressed.

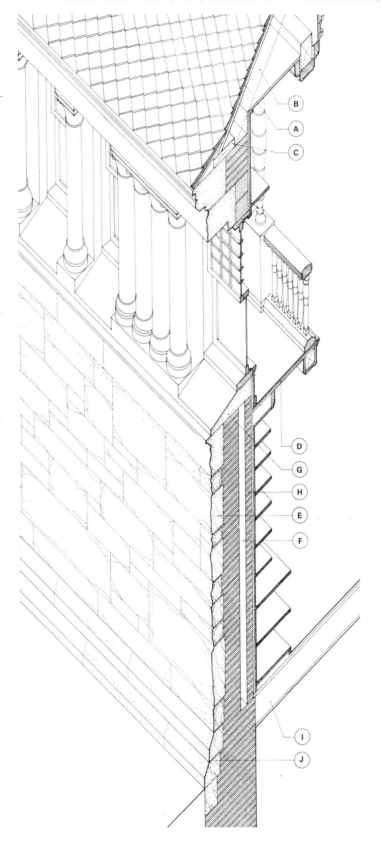

CHAPTER 1 Monolithic bearing walls – masonry, stone, concrete

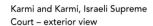

Karmi and Karmi, Israeli Supreme Court – exterior view

components within the wall assembly. The use of expensive stone where durability, weather protection, and appearance dictated it reminds us of the traditional Roman reinforced masonry construction technique where rubble masonry and concrete were used behind more expensive brick, in a "wet set" masonry wall.

Modern examples of monolithic stone masonry include the Israeli Supreme Court, completed in 1992. The product of a competition won by Ada Karmi-Melamede and Ram Karmi in 1986, the Supreme Court building is located on a commanding site in Jerusalem, adjacent to the Knesset, and bridges between the Mediterranean Sea to the west and the desert to the east. As a symbol of both modern Israel and its historical and cultural roots, the stone building offers a backdrop to the theater of traditional Hebrew law. Organized to create a public space representing the relationships between the executive, legislative, and judicial branches of the government, the Supreme Court building incorporated courtyards and light wells throughout the building to temper the strong daylight. Veneer stone on poured-in-place concrete was used in a variety of textures (see Figure 1.55) to play with light and shadow in this relatively hot, dry climate. Figure 1.56 shows stone details. Arcades and passages along the interior courtyards provided shelter from the daytime sun, as shown in Figures 1.57 and 1.58.

The Dominus Winery, Yountville, California, was designed by Jacques Herzog and Pierre de Meuron, and constructed in 1997 using stone captured in a steel mesh cage (see Figure 1.59). This system, commonly used in civil engineering for erosion control and earthworks, normally allows groundwater to filter out between rubble masonry, or "riprap" stone. In the elegant winery, these galvanized steel gabions, or cages, enclose basalt stone from a nearby canyon and are used in combination with tilt-up concrete construction to form

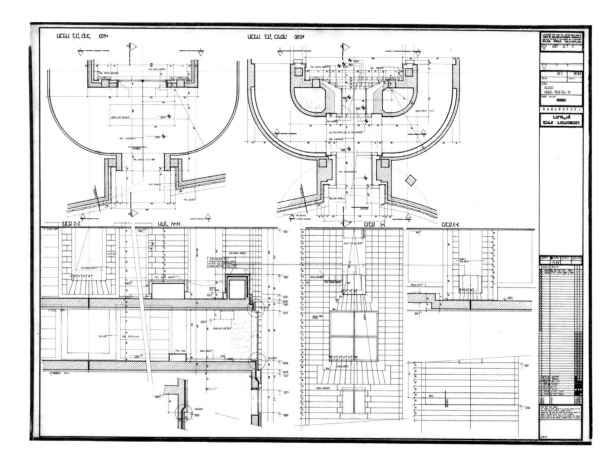

▲ Figure 1.56

Karmi and Karmi, Israeli Supreme Court – detail drawing

the enclosure for the building. In the Napa Valley climate the stone is exposed to the heat of the daytime sun, absorbs the solar radiation, and re-radiates the heat to the interior slowly over the cold nights in a daily cycle. Careful selection of rock sizes was used to filter daylight yet permit glimpses of the vines outside (see Figure 1.60). The thick masonry in its cages with great thermal capacity shade the concrete walls of the building inside, creating an above-ground thermally buffered shelter for winery functions – the tank room, the wine cellar, and the storeroom. The long, low building is given a textured, yet minimalist breathable skin by the open masonry assembly, reflecting the horizontality of the vines outside (see Figure 1.61).

Concrete

Concrete is made of three major components: Portland cement (lime, iron, silica, and aluminum); aggregate (coarse and fine); and water. The mixture gives off heat during its hardening, or "curing" phase, which is a process of hydration as the bonds form between these component materials. There are eight types of concrete as defined by the American Society for Testing and Materials,

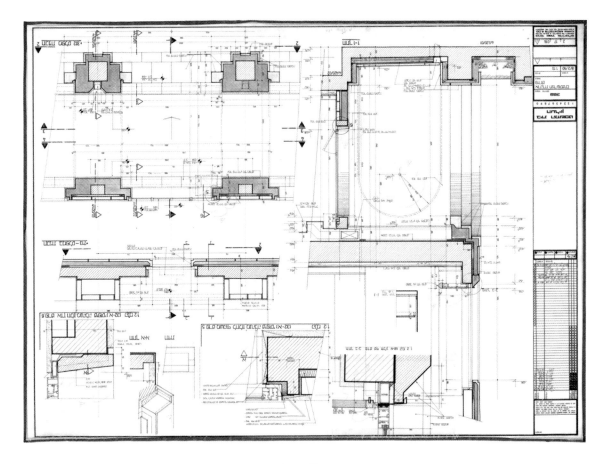

▲ Figure 1.57

Karmi and Karmi, Israeli Supreme Court – detail drawing

where distinctions are made between concrete for different uses, strengths, and exposures. Lighter aggregates can be used to produce lightweight concrete, for example, a non-structural concrete mixed with vermiculite aggregate, which gives lightness and good insulating properties. The mixture is poured, or "placed", into formwork molds. Form liners can be used to give texture and finishes, in addition to handwork, which becomes expensive as the cost of labor goes up; but historically concrete, like stone, could be bush hammered, sand blasted, washed with acid or otherwise further doctored. Concrete can be cast on the job site, traditionally, or pre-molded off-site, and called "precast".

Like concrete masonry units, precast concrete systems are manufactured off-site. Concrete can be poured into molds and shaped into components that are manufactured under controlled conditions in a plant, rather than relying on field conditions to determine pour schedule and curing conditions. The elements of a precast wall, however, tend to be much larger than either a brick (set with one hand) or a concrete masonry unit (set with two hands), but rather have to be lifted into place by crane, most commonly. A load-bearing monolithic precast concrete example is the Wesleyan University Creative Arts Center by Kevin Roche John Dinkeloo and Associates, 1965–1973 (see Figure 1.62). This project for a college campus in Connecticut was designed in bearing wall

▶ Figure 1.58

Karmi and Karmi, Israeli Supreme Court – courtyard view

construction, detailed as a precast concrete kit of parts. Individual components could then be made in a repetitive, economical way, but with a higher degree of quality control than site-poured work. Panel sizes for larger-scale prefabricated construction including precast are limited, however, by method of transport and weight in lifting. Elements can be poured and engineered to be up to 50 feet long, for example, but bridge and highway limits, as well as site crane lifting limitations, can prevent components from getting much larger without significant cost impact.

▲ Figure 1.59

Herzog & de Meuron, Dominus Winery

◀ Figure 1.60

Herzog & de Meuron, Dominus Winery

▲ Figure 1.61

Herzog & de Meuron, Dominus Winery

Site-cast, or Cast-in-place Concrete, unlike precast, is not limited by transport or jointing. In most cases, the design is really limited only by the formwork and how it can be shaped. In traditional concrete work, formwork meant carpentry, wood forms having to be built first before the concrete itself could be placed. Malleable, sculptural building enclosures can be created so long as there is imagination, skill, and money enough to create the wood mold impression that must first be built. Early innovations in industrial construction took advantage of the tensile strength inherent in steel and the ability to bond well onto poured concrete, which like masonry is very good at taking and transferring loads in compression. By enclosing steel bars within the thickness of concrete, it is possible to minimize and prevent cracking in the cover material when bending might otherwise cause structural failure.

Early examples of warehouse and factory uses exploited the close thermal expansion rates of concrete and steel. Reinforced concrete was pioneered in the late 1800s as "béton armé" by François Hennebique, initially as a fireproofing of steel by encasing it in concrete. The resulting assembly of beams and spanning slabs, integral to the column structure of the building, was continued and brought into the architectural realm by architects such as Auguste Perret, and then Le Corbusier. At the church in Ronchamp, through later frame-articulated projects such as the Convent of La Tourette, or the Unité d'Habitation at Marseille (Figure 1.63), Le Corbusier took advantage of reinforcing steel to allow greater freedom, bigger spans and cantilevers since it performs well in tension, while the concrete is very good for accepting and distributing structural loads in compression. The bond between steel and concrete is very strong, and they share similar coefficients of expansion, therefore the only drawback is that the permeability of concrete can lead to the oxidation or rusting of the steel, if the concrete is not protected from water and rain.

Other innovative examples of monolithic reinforced concrete construction, for their time, include the works of structural engineers Felix Candela, Eduardo Torroja, and Pier Luigi Nervi. Their projects reflect forms whose shapes were developed as the understanding of structural loads and pathways became identifiable and formwork technology or hand craft permitted. Torroja's experiments from the 1950s also explored minimalism in structural expression, using thin concrete shells whose form matched the efficient flow of structural loads. Likewise, Pier Luigi Nervi's Sports Palace projects, in Rome, 1957, located reinforcement steel in a filigree of ribs that express the flow of tensile forces in the sports hall roof, cast as coffered repetitive shapes into a ring of concentric geometries (see Figure 1.64).

At the Los Manantiales restaurant in Xochimilco, Mexico City, Mexico, completed in 1958, Felix Candela cast very thin, elegant shells into a hyperbolic paraboloid form. Figure 1.65 shows an interior view of the restaurant. Like his other thin shell studies and projects, the reinforced concrete took the form of the compressive stresses, allowing the concrete to perform extremely efficiently with minimum thicknesses of material. Concrete coverage of the rebar and construction tolerances become as important as load transfer in these conditions. Figure 1.66, which shows the building under construction, illustrates the elaborate formwork required to achieve such flights of structural efficiency. Quality and cost of wood formwork, which is then completely removed from

▶ Figure 1.62

KRJDA, Wesleyan University Creative Arts Center

▲ Figure 1.63

Le Corbusier, Unité Marseille

▲ Figure 1.64

Pier Luigi Nervi, Little Sports Palace/Palazzetto dello Sport, Rome, interior view looking up at coffered concrete dome

▶ Figure 1.65

Felix Candela, "Los Manantiales" Restaurant, Xochimilco, Mexico City, Mexico, interior view

CHAPTER 1 Monolithic bearing walls – masonry, stone, concrete

◀ Figure 1.66

Felix Candela, "Los Manantiales" Restaurant, Xochimilco, Mexico City, Mexico, construction of concrete shell

▲ Figure 1.67

Felix Candela, "Los Manantiales" Restaurant, Xochimilco, Mexico City, Mexico, completed shell, overall view

the finished work, is not insignificant. A very dry mix of concrete is shown being packed onto the formwork, with the reinforcing mesh of steel captured within the shell thickness. Upon completion, the profile of the newly finished structure is clearly a pure reflection of the generating geometry (see Figures 1.67 and 1.68).

More recent examples of reinforced concrete construction include Herzog & de Meuron's Library at the Technical College of Eberswalde, from 1999 (see Figures 1.69 and 1.70). There, artist Thomas Ruff's imagery was applied like photo-transfers to the concrete formwork, onto very simple poured-in-place concrete walls with punched openings. Design considerations for conventional site-cast concrete include formwork, surface finish (form liners can be used to give texture to the finished concrete, if desired), and through-ties that hold the two sides of the formwork at the desired set dimension, counteracting outward pressure from the weight of the wet concrete. The simplest wall forms can be boards or plywood, braced by struts and held together with form ties. Concrete is placed, or poured, into the formwork around any reinforcing bars. Joints between pours, called cold joints, can be noticeable in any concrete finish exposed to view.

Control joints must be placed frequently enough to provide a controlled location for cracking to occur, or uncontrolled cracks can become problematic to the integrity of the entire assembly. Joint placement concerns are also important to consider at openings for doors or windows. If notched weak lines are not created, to force cracking into predetermined paths, the resulting cracks can also be unsightly as well as leading to long-term damage. It can be important, then, to consider architectural features and design elements to conceal these conditions or otherwise take account of inevitable cracks from shrinking of concrete and minor deflections from live lateral loads.

◄ Figure 1.68

Felix Candela, "Los Manantiales" Restaurant, Xochimilco, Mexico City, Mexico, finished shell

◄ Figure 1.69

Herzog & de Meuron, Eberswalde library, concrete imprinted with art by Thomas Ruff

▶ Figure 1.70

Herzog & de Meuron, Eberswalde library, view of entry

Insulation placement options in concrete walls can obviously include insulation placed on the interior face with vapor barrier on the warm side. However, conducting condensation in a path to weep out to the exterior can be difficult to achieve with insulation adhered to the interior face of the wall. If insulation is placed within the concrete, bonding can be a concern. The core insulation line must be broken sufficiently to ensure load transfer from the visible exterior side of the wall back through the insulation to the interior side of the concrete wall. Associated difficulties may make consideration of the monolithic bearing wall unfeasible. If insulation is placed at the exterior, then protection of the insulation itself from damage from the elements may incur an additional exterior layer of

◀ Figure 1.71

Tadao Ando, open-air chapel, UNESCO Headquarters, Paris, France, interior

◀ Figure 1.72

Tadao Ando, open-air chapel, UNESCO Headquarters, Paris, France, interior

material, thus turning the concrete into the backup structural portion of a cavity wall.

Concerns regarding insulation placement are completely relieved in the open-air meditation chapel at the UNESCO headquarters in Paris, France, designed by Tadao Ando. Completed in 1995, this tiny reinforced concrete cylinder is set within a Japanese garden constructed to celebrate the fiftieth anniversary of the founding of UNESCO with a nondenominational space for people from many cultures and religions. The simple form reveals its method of construction with the six form tie indentations, which remain after the ties are snapped off. The cold joint marks a mid-height datum line on the drum, interior and exterior, and the interior is lit by the simple portal and the raking light from a perimeter gap around the entire roof, simply supported at four points onto the side walls of the cylinder (see Figures 1.71 and 1.72).

Chapter 2
Load-bearing cavity walls

Cavity wall construction consists of a bearing wall assembly where a gap is created between the outer surface, exposed to view from the exterior of the building, and an inner plane of construction. The interior half of the assembly is typically the portion that provides structural integrity to the composite assembly. The outer surface is exposed to the elements and is therefore the primary weather barrier. Its nature determines the initial watertightness of the skin. The outermost surface of the outer wythe, in the case of a masonry cavity wall, forms the primary line of defense against water and air infiltration into the assembly. By creating a two-part composite wall, the inner half is protected from the immediate elements (see Figure 2.1). This inherently lays out a series of relationships

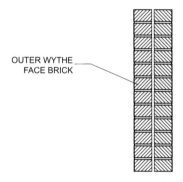

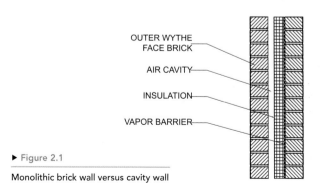

▶ Figure 2.1

Monolithic brick wall versus cavity wall

fundamental to the principles behind dual-seal exterior enclosures systems, in general, and leads to the pressure-equalized rainscreen that carries through into non-load-bearing curtain wall and cladding design.

Weather and rainwater

A "solid" or monolithic masonry wall as we studied in the previous chapter, when used as the only barrier between exterior weather and the tempered interior environment, has a number of shortcomings – there is inevitably the potential for water penetration since masonry is naturally somewhat porous and, being assembled from many small units, tends to have joints which are susceptible to failure; it is reliant on masonry joints for performance since the mortar and surface seals must resist water infiltration while also accommodating thermal expansion; there is poor thermal performance since there is limited insulation in most masonry types, and they have relatively poor insulation values inherently since resistance to heat flow is not great.

The typical monolithic wall, then, is essentially a single-seal system with only one "line of defense" against the exterior environment. Any deficiencies in workmanship or material would result in failure of the skin system. Absorption of moisture or rainwater penetration into the solid wall results in a transfer of the water through the wall itself in either water or vapor form. When traditionally whitewashed, or painted with a mixture of ground-up lime rocks, the interior face of such a wall could breathe as easily as the heavy masonry wall itself, letting in exterior ambient moisture, without restriction. Where used in relatively dry climates, or in regions where cycles of rain alternate sufficiently with dry weather, such a wall can dry out sufficiently to avoid problems of mold, damp, or other concerns.

The introduction of an air gap between the exterior and interior surfaces of the enclosure wall, even if uninsulated, provides the opportunity to control water penetration through the exterior portion of the assembly, and potentially a way to drain out water or condensation between inner and outer walls of the cavity assembly. Examples of early cavity wall projects include libraries such as the Ames Library, Boston, or McKim, Mead & White's 1906 Morgan Library (see Figure 2.2). To store books and protect against deterioration from moisture, these projects introduced the split masonry wall to prevent any dampness from damaging the precious rare books the buildings were designed to contain (see Figure 2.3).

In the Morgan Library, a formal, classical vocabulary of architecture was used to evoke the importance and reference classical learning. The building was actually constructed with an early form of cavity wall construction (see Figure 2.4). Thick stone walls were faced at the exterior with Vermont white marble set with very tight tolerances, so as to appear dry set, but were actually assembled with mortar and lead sheets for flashing and waterproofing. The thick stone walls form a veneer, exposed to the exterior, but rely on a less costly brick backup wall for structure. In the example of the Morgan Library, the cavity was also coated with waterproofing on the outer surface of the inner brick wall, to

▶ Figure 2.2

McKim, Mead & White, Morgan Library, exterior view

▶ Figure 2.3

McKim, Mead & White, Morgan Library, interior view

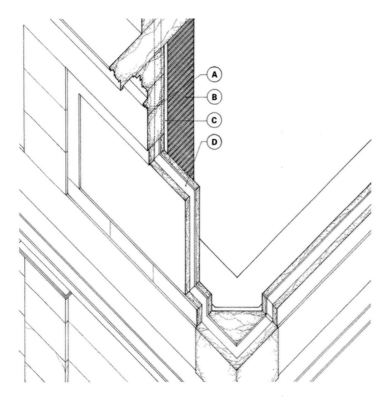

◀ Figure 2.4

McKim, Mead & White, Morgan Library, axonometric showing proto-cavity wall. A – Horizontal joint in white Vermont marble wall. All joints are built to zero tolerance and filled with thin sheets of lead. B – Brick wall. C – Air space. An air space is provided for waterproofing. The marble is anchored to the brick. D – Slot. All horizontal and vertical joints were given pockets, which were then filled with Portland cement. Although the joints appear to have been set dry (without mortar), they were not. The masonry contractor felt that this type of marble was too irregular in consistency for "perfect" joints

further protect the valuable books stored inside against moisture. Metal ties anchored the veneer stone to the backup brick wall, much as in modern cavity wall construction.

The introduction of an air gap between outer and inner wythes of cavity wall construction also creates an opportunity to increase the insulation value of the overall assembly. Though there is air in the cavity, if it is not captured and sealed, the insulating effect is limited. By combining an actual insulation material (lofty light materials, often, with the ability to hold air within their dimension) with the cavity, it becomes possible to significantly improve the overall thermal performance of the exterior skin. Figure 2.5 illustrates how a cavity wall with insulation in the cavity can significantly change the thermal gradient as compared with a monolithic, uninsulated wall. Where the outside temperature is roughly 0 degrees Fahrenheit and the interior temperature is a comfortable 70 degrees Fahrenheit, the temperature gradient if measured at any point across the wall is a straight slope increasing in temperature the closer we measure to the interior. By introducing insulation within the cavity, the temperature gradient at the outer face of the inner wythe can be increased significantly. Depending on atmospheric humidity, this can mean that the dew point, the temperature at which moisture in air of a given temperature precipitates into water, is shifted into the cavity, as shown in Figure 2.6. If the condensation occurs within the cavity it becomes possible to introduce a waterproof surface to conduct water droplets away from the interior, through a weep hole in the veneer masonry, in a controlled fashion.

▶ Figure 2.5

Temperature gradient (in degrees Fahrenheit) across monolithic brick wall versus cavity wall

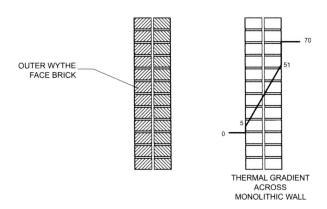

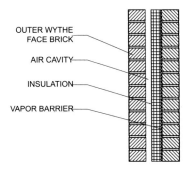
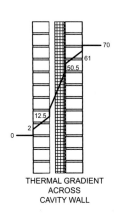

▶ Figure 2.6

Dew point chart showing when condensation occurs

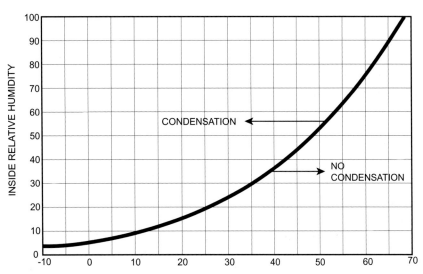

CHAPTER 2 Load-bearing cavity walls

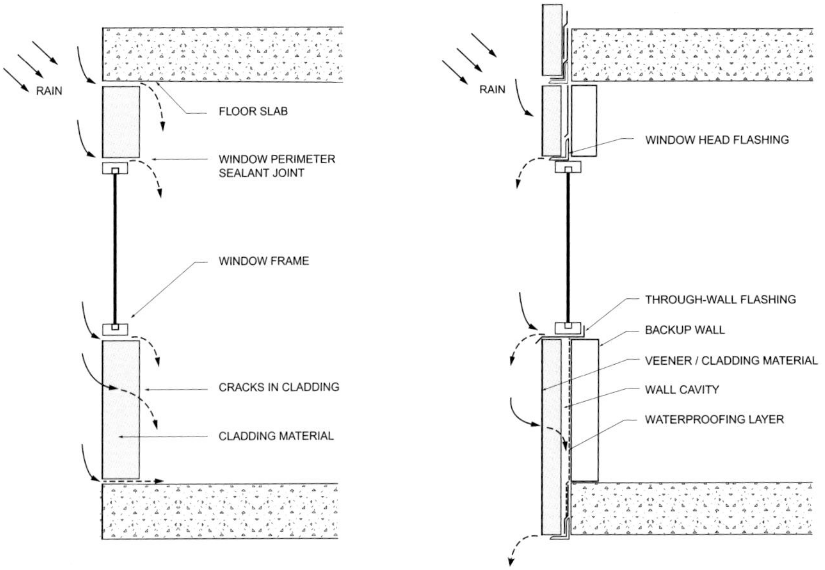

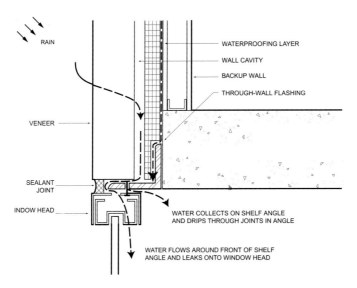

◀ Figure 2.7

Water infiltration vulnerabilities

◀ Figure 2.8

Where to flash

Water infiltration vulnerabilities occur at every opening and gap in the exterior skin. Besides water collecting in and around each infill condition and potentially flowing in through window openings in the masonry, the basic porosity of a veneer can allow rain to seep into the brick or stone (see Figure 2.7). Using waterproof materials overlapped and ideally sloped will allow gravity to drain the moisture to weep holes in a controlled manner (see Figure 2.8).

As we also saw in the previous chapter, monolithic masonry walls consisting of stacked load-bearing elements are able to span greater heights if they are reinforced with steel ties to increase strength against tension and shear forces. In modern masonry construction, prefabricated galvanized steel trusses laid horizontally within the depth of bed joints provide much more capacity than masonry elements alone. These horizontal members take the tensile forces to prevent formation of vertical cracks. Vertical steel reinforcing bars resist forces and can help eliminate horizontal cracking. Together, these types of reinforcing can be used to increase the height limit of masonry bearing walls without necessarily requiring thickening of the wall at the base like the Monadnock Building (Figures 1.27 and 1.28).

To restrain the outer veneer masonry, ties are also used to provide lateral stability for veneer masonry. In regions where seismic concerns do not need to be addressed, simple galvanized or stainless steel ties can be used, set within the bed joint of both the outer veneer masonry and the inner backup structure. Conveniently, many brick sizes have been standardized in North America such that three courses of brick align with one course of concrete masonry units, in the vertical dimension. Brick ties can thus be inserted into the horizontal bed joints at these coplanar conditions. Seismic ties, consisting of two-part heavy steel wire in a variety of proprietary configurations, allow for some slip between veneer brick and backup CMU planes in the event of an earthquake. These ties can typically be thin enough to avoid creating a significant thermal break between the outer and inner brick, and the cavity insulation can easily be placed around or between reinforcement ties.

Masonry also has the ability to retain heat/coolness if a sufficient quantity is used, as it has good thermal mass. In hot, dry climates, for instance, a heavy masonry wall can be heated slowly by the sun, over the course of the day, and then release the heat at night. Sustained hot and cold seasons, however, need wall assemblies with greater thermal resistance than can be achieved with thin masonry construction. Higher R-value, or greater insulating capacity, retains heat or cold better within an airtight enclosure. Thermal capacity in stone masonry, for instance where used in medieval churches and cathedrals, allows sunlight to heat the thick walls, which then radiate the heat slowly enough because of the great thermal mass on a cycle offset from the daily cycle of daytime sun exposure/nighttime cool temperatures. In summer, the cool night air allows the walls to dissipate the heat absorbed during the day, and conversely will chill the inside air for many hours during the day. Thick foundation walls can further capture the temperature of the earth, if deep enough below the surface to eliminate weather and climate effects, tempering the air and objects stored in habitable spaces, like wine in the case of cellars or caves, or the example we saw in Chapter 1, Figure 1.17.

The thermal capacity inherent in heavy masonry and concrete was exploited in Douglas Kelbaugh's house at 70 Pine Street, in Princeton, New Jersey, completed in 1975 after the oil crisis of 1973. To minimize heating costs, a two-storey-high poured-in-place concrete wall was constructed, painted black, and covered with a single-glazing system (see Figure 2.9). Oriented directly south to capture as much solar radiation as possible, vent openings placed at floor level allowed cool air to enter the cavity, where it would be heated in the cavity, and rise upward through convection forces (see Figure 2.10). Outlets under the

▲ Figure 2.9

Douglas Kelbaugh, House at 70 Pine Street, Princeton, NJ

ceiling allowed warm air to circulate into the rooms on the north side of the wall, creating convection currents within the habitable spaces. This passive solar house also incorporated carefully placed windows and skylights to ensure that cross ventilation would provide cooling breezes for the more temperate seasons.

One mechanism that provides opportunities to implement these details is to create a cavity behind a rainscreen surface. Small gaps in the rainscreen then create a condition where the pressure inside the cavity achieves equilibrium with the exterior air pressures. This is not terribly significant at grade, perhaps, but the principle extends to very tall construction where exterior suctions and pressures can become significant, and the rainscreen principle reduces the stresses on the system at those peak conditions.

With a rainscreen, most of the water is screened by the exterior surface from reaching the interior, which is reduced to acting as a barrier for air infiltration into the building.

A single seal at the exterior face is significantly less effective without a rainscreen or pressure-equalized cavity since any tiny breach becomes problematic (see Figure 2.13). Monolithic barrier systems must rely entirely on the exterior

▶ Figure 2.10

Douglas Kelbaugh, House at 70 Pine Street, Princeton, NJ

How water penetrates

Cavity walls and moisture control

Cavity wall systems developed by splitting two wythes of a masonry wall improves thermal performance slightly by capturing air, which is a very good insulator. The addition of insulation in the cavity within the wall, against one side of the masonry or other, and inside the cores of CMU, if used as a backup wall, are further improvements. Temperature gradient changes depending on the thermal resistance of the layers in a wall assembly. As an example, Figure 2.5 shows how the consistent temperature gradient in a monolithic wall increases with the introduction of a cavity. The addition of insulation with a vapor barrier on the warm side can improve the air and condensation performance of a cavity wall even further.

Condensation occurs when moist air is cooled to the dew point temperature. Air can hold only so much water vapor before it turns into liquid form. The dew point chart illustrates that if the temperature is high, air may be able to hold greater quantities of water vapor before condensing (see Figure 2.6). When the hot, moist air touches a cool surface at or below the dew point, liquid water forms and will travel wherever gravity leads it. This can lead to significant water formation, spalling or delaminating of cladding materials, corrosion of anchors and building steel, or deterioration of insulation and interior finishes. By placing an impermeable material (metal or man-made rubber) on the warm side of insulation within a wall assembly, the condensation formed can be controlled and directed to the exterior safely. Venting the exterior will allow the moisture to evaporate (in a cold climate, this means venting the cold side). The insulation selected must be able to tolerate some exposure to moisture, typically rigid polisocyanurate or possibly mineral wool, rather than a type that would be vulnerable to deterioration in a wet environment, such as fiberglass. The vapor barrier located at the interior face indicates that this assembly is to be installed in a geographically heating-dominated, colder climate, since the interior is likely to be warmer most of the year, than the exterior.

It is important to avoid trapping moisture within a wall since decay from exposure to water droplets can deteriorate materials within the exterior assembly. Depending on design, construction, and environment, the careful placement of the vapor barrier can address many of these concerns. In a humid, tropical region, the vapor barrier might be placed at the exterior face, if the outdoor temperature and humidity is consistently expected to be a greater concern than an air-conditioned, inhabited interior. In the colder climate, the vapor barrier would need to be placed on the interior face of the insulation, for the opposite reasons. It then becomes simpler to remember that the vapor barrier should be located at the warm side of the insulation (see Figure 2.11). If by accident two vapor barriers were to be installed, the risk of creating a region within the wall assembly of trapped condensation from

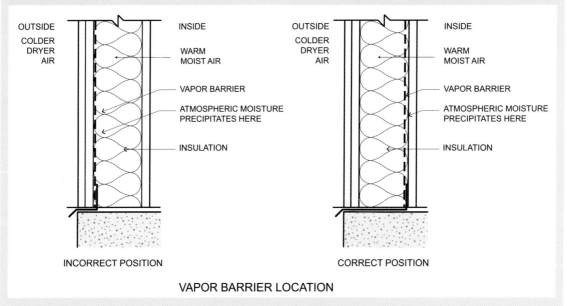

▲ Figure 2.11
Always locate the vapor barrier at the warm side of the insulation

How water penetrates *continued*

water vapor in the cavity becomes a concern. Formation of mold has also become a significant problem in some regions, so avoiding a double vapor barrier can eliminate trapped moisture conditions.

Since whitewash paint on clay brick or plaster walls is no longer as common as it once was, finish selection can also become a point of concern for vapor control. The popularity of vinyl wallpaper in hotels and motels, when they first became available in the 1970s and 1980s, for example, led to problems in subtropical areas of the United States. The vinyl sheets adhered continuously around the interior face of insulated perimeter walls inadvertently created a second vapor barrier creating mold on the wrong side. When removed, mold was often discovered hidden on the surfaces since they formed a moisture-proof membrane on the cold side of the insulated exterior walls, and the moisture had no pathway to vent back out to the exterior or naturally dry out. It is important to select a sufficiently porous outer veneer allowed to dry out, between rains, or otherwise design an assembly that can incorporate mechanisms to discourage water from entering into the system, as well as build in elements to drain out any moisture or water that forms within the assembly, and conduct it out in a controlled manner.

Figure 2.12 shows how water penetrates. Water must be present at the outer face of the assembly:

1. There must be an opening for water to move through.
2. There must be a force to propel it through the opening. These forces can include:
    a. Gravity (which can be eliminated by building in an incline against the likeliest water flow patterns)
    b. Momentum (horizontally wind-driven water can be prevented from entering an assembly using a baffle, or labyrinth)
    c. Surface tension (water clings to the materials and is drawn into the system surface tension. This can be prevented with a "drip", or gap large enough to prevent bridging)
    d. Capillary action (water droplets are drawn into the system by a crack, tube, or gap that allows differential pressures to bring moisture further into the assembly)
    e. Air flow (control of convection and air currents into the opening and cavity beyond, including careful consideration of relative pressures)

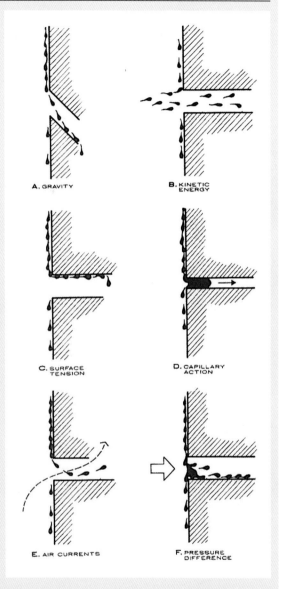

▲ Figure 2.12

"Forces acting to move water through an opening"

CHAPTER 2 Load-bearing cavity walls

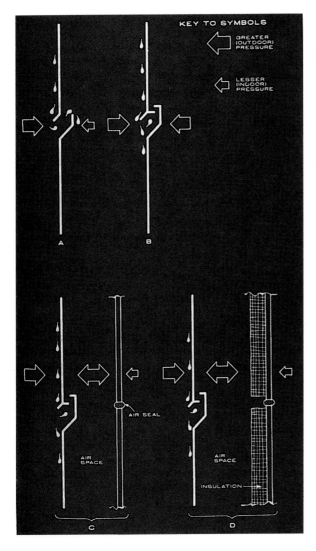

◀ Figure 2.13

Rainscreen pressure equalization: correct configuration of flashing or baffles can be used to limit suction of moisture in through the skin. By ensuring that internal cavity pressures are equalized to the exterior atmosphere, water infiltration can be even better controlled. This is achieved by creation of an airtight secondary surface within the cavity. Even if indoor air pressures are less than outside wind pressures on the building, the cavity can achieve equal pressure to the wind by means of small weep holes and the resulting assembly can best screen rain from penetrating. The rainscreen also presents an opportunity for insulation to be placed within the cavity, in conjunction with a vapor barrier at the warm face of the insulation.

cladding and surface seals, with no tolerance for construction variations or defects in workmanship, degradation of materials, lack of maintenance on seals at joints, and thermal expansion/contraction breaks over time. Dual systems with a rain- or water-shedding primary seal and secondary air barrier line of defense can provide much better performance with essentially the same materials. The simple addition of a cavity means the exterior veneer sheds most of the water and limits the quantity of water reaching the interior surface. The interior line of defense becomes secondary, providing a water- and airtight backup system to collect water that does enter the system and drain it to the exterior.

The cavity in a rainscreen system can be compartmentalized to reduce any chimney effect with the same flashing that also conducts water to the exterior. Flexible joint sealants can be used to close expansion/seismic/movement/construction joints and also function to define the exterior water seal. Gaps in the exterior seal for weeping out drained moisture provide a dual function as the means by which pressure can be equalized into the cavity. Exposure to sunlight, UV, water, temperature changes, and movements can degrade these sealants, but silicones have been developed to withstand these various forces and still maintain elasticity.

Proper flashing installation versus improper flashing installation

Masonry and exterior veneer is a porous surface, especially when mortar joints fail. Cavity prevents water from reaching the interior surface, and if pressure is equalized to the exterior wind, allows water to drain down and out by gravity. Flashing must, however, be installed at and around openings to ensure that the water is conducted outwards properly. Flashing must also be impervious to water and maintain continuity (see Figure 2.7). Lightweight materials can be punctured or torn during the construction process. Historically, thin metals were used exclusively for flashing – copper, lead, stainless steel and its precursors. More recently, membranes and fabrics with bituminous coatings have become prevalent. Where exposed to sunlight, it is important to ensure that the material selected will not degrade and can form an exposed drip edge, ideally extending past the face of wall for maximum protection, like the brim of a baseball hat.

In a brick cavity wall, it is ideal to channel water to the exterior roughly 18–24" on center, either in plastic or metal tubes laid in the mortar joints, or historically with oiled rope or simply open gaps in the mortar. Various proprietary weep products are commonly used to ensure that passageways for water from the cavity to the exterior are maintained over time. In all cases, the flashing must lap positively and shingle over the cavity at openings such as windows and doors, and lead to these weep points.

Mortar

Joints are the weakest link and therefore the most likely places for unwanted water penetration to occur. Water running down the face of an exterior wall is liable to accumulate at joints and, especially if not allowed to dry between wetting then exposed to freezing conditions, can cause spalling and cracking, flaking off mortar and creating areas of leakage, loosening joints, and beginning a downward spiral of deterioration. To prevent this, mortar should be selected to be suitably water-resistant in composition, well filled and compacted into a concave profile or V-joint to ensure adhesion to adjacent masonry.

Notable examples

An interesting example of brick used in modern architecture is the Exeter Library by Louis Kahn, 1965–1972 (see Figure 2.14). The project incorporated brick cavity wall construction with exposed brick at both interior and exterior

▲ Figure 2.14

Louis Kahn, Exeter Library, exterior view

surfaces. Brick was used structurally, spanning window openings with flat arches, in a three-wythe-thick veneer layer. The interior ring of structure was formed of poured-in-place concrete, with a cavity wall outer concentric ring. As shown in Figure 2.15, true brick flat arches spanned the large openings in the outer brick wall, without steel-concealed lintels. For this project, Kahn designed an outer layer of 12" overall, consisting of a 4" brick, 8" CMU common bond pattern and an inner wythe of 4" nominal running bond, with 1½" rigid insulation, a moisture-resistant product, within the cavity (see Figure 2.16).

Alvar Aalto's Säynätsalo Town Hall, Finland, 1949–1952,[1] used brick and timber to delineate a composition of buildings clustered around a raised central courtyard in the heart of a new town planned for the Finnish forestry products company Enzo-Gutzeit. The winner of an invited competition, Aalto assembled the administration building, town library, offices, retail spaces, and apartments around the raised central plaza with the Council Chamber building cantilevered into the wooded landscape, with concealed beams and support lintels for the brick (see Figure 2.17). In this building, cavity wall construction was used to lighten the cantilever loads. Because of the nature of the environment, double-paned insulating glass units were required as well as cork insulation within the cavity.

Like the cantilevered Council Chamber, Alvaro Siza's Center for Contemporary Arts, 1995, used a cavity wall at the exterior enclosure for the

▶ Figure 2.15

Louis Kahn, Exeter Library

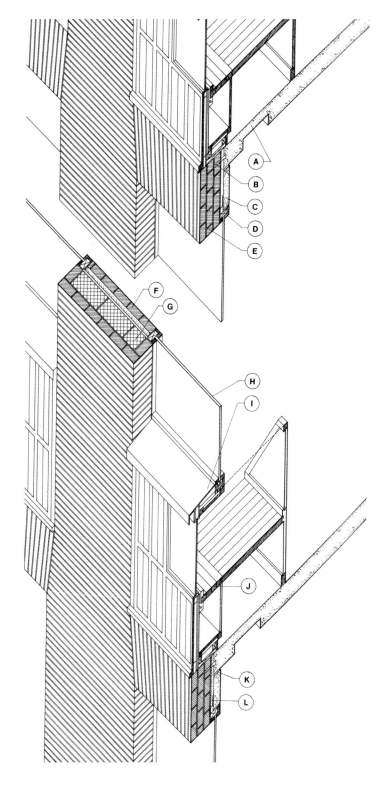

◀ Figure 2.16

Louis Kahn, Exeter Library. A – concrete slab with slot for light fixture. B – anchors 12″ oc to tie arch back to continuous dovetail anchor slot in concrete slab. C – sandstone lintel. D – teak and oak window frame with flashing to divert water out of cavity. E – brick arch. F – inner wall of 4″ brick with cavity filled with 1½″ rigid insulation fastened to ¾ galvanized furring channels. G – brick-faced concrete masonry with headers every fifth course. H – 1″ insulating glass. I – stainless steel sill on wood frame to drain water off horizontal surface. J – carrel of 1″ teak boards, 1″ rigid insulation, ¾″ oak plywood, and ½″ insulating glass. K – compressible filler to ensure that the slab does not bear on the lintel. L – flashing to drain water from davity to exterior

▶ Figure 2.17

Alvar Aalto, Säynätsalo Town Hall

▲ Figure 2.18

Alvaro Siza, Center for Contemporary Arts

small museum building in Santiago de Compostela. In this example, however, the exterior veneer was granite cladding, supported by concrete backup on a dramatic, long, steel lintel. Figure 2.18 shows that a large steel angle was used to support the weight of the stone veneer above the wide opening. Within the cavity wall assembly, insulation and warm-side vapor barrier was lapped over the steel lintel draining water out to protect from condensation, which can

▲ Figure 2.19

Alvaro Siza, Center for Contemporary Arts

be a significant concern in museum construction and other interior environments requiring high humidity – in the case of museums, to protect the valuable artwork within. The window is clearly set deep within the opening as well, and given sufficient space to be set into the opening with room for tolerances to achieve a clean finished appearance (see Figure 2.19).

Chapter 3
Light frame construction

The last of the three bearing wall systems we will consider is light frame construction. This method of construction is still commonly used in places where wood and framing materials are readily available and where, at least historically, extensive forests provided plentiful sources of timber at a time when centralized milling into lumber and the industrialized equipment to process them became commonplace, resulting in an inexpensive and plentiful supply of standardized lumber. The advantages of this type of construction are that it can be erected swiftly, is easily modified and flexible enough as a system to be used to achieve many forms and shapes, and can be constructed with relatively unskilled labor. Drawbacks of wood construction are that, as a natural material, it burns easily, decays when wet, experiences significant expansion and contraction, and unlike the two previous systems we have examined, needs additional finish, possibly the work of other trades. Metal studs have further developed light frame profiles for use in situations where combustible materials are inappropriate, like multi-family housing or commercial interiors, while still preserving the lightweight, flexible nature of the construction type. In another direction, industrialized production of laminated wood, consisting of wood layers or particles bound together with glues or resins, has developed another facet of structural use of wood in construction.

The idea of shelter in the history of Western architecture, rooted in the temperate climate zones where forests prevailed, is tied to a perhaps romanticized association with the primitive hut. From Laugier's image of the Vitruvian primitive hut, in the frontispiece of his *Essay on Architecture* of 1755, the notion of shelter was rooted in a structure formed from vertical wood posts and horizontal wood beams (see Figure 3.1). Laugier's argument for clarity in expression was perhaps a romanticized view, looking back to prehistoric times when posts had to be functional and rafters for supporting the roof had to be triangulated to provide balance. The visual emblem of shelter was created by the triangular pediment. The clear articulation of column and lintel, though magically suspended in this idyllic portrayal, provides a conceptual source and visual reference to the realities of construction detailing in wood and timber.

At the other extreme from high culture discussions of authenticity in expression and articulation of the visual language of architecture, wood used in vernacular construction was somewhat beneath the notice of architects until after the early modern era. The material was available, plentiful, and relatively easy to shape into useful components. Principles of assembly formed a basis for the vocabulary and method of framed assemblies later transferred to more

◀ Figure 3.1

Laugier, primitive hut

▶ Figure 3.2

Log cabin, Harrells, Sampon, NC

durable, less combustible materials than wood. Especially with developments in larger-scale cutting and milling, beginning with early North American settlers and culminating in the mass production of lumber still prevalent today, the use of light frame construction is interesting to consider within the spectrum of load-bearing, yet material efficient, envelope systems.

Log cabin construction from early North American examples is perhaps the clearest example of stacked wood construction, with only the simplest intersection notch required to achieve stability in two directions (see Figure 3.2). The weight of the massive walls served to prevent overturning and ensure that the interlock between logs remained intact. This crude method of building was possible where large trees of fairly regular diameter were plentiful and skilled labor unavailable. Gaps in logs were filled with mud, clay, or other pliable material that would keep the wind and rain out, and easily renewable when wood swelled, shrank, or otherwise might disturb the dried filler (see Figures 3.3 and 3.4). These simple stacked walls were constructed in places where wood was plentiful, time and skilled labor might be lacking, and the need for shelter was critical to survival.

Unlike the log cabins of the American frontier, societies and cultures where wood joinery was developed to its highest degree in post-and-beam or post-and-lintel heavy framing required skilled joinery and carpentry skills. Used in boatbuilding, fine woodworking for furniture, and in architecture and construction since early human history, connections between two wood members had to be carefully prepared in order to make a secure joint, able to resist overturning

▲ Figure 3.3

Log cabin, Joker Mine

◀ Figure 3.4

Log cabin – detail of construction

▲ Figure 3.5

Mortise-and-tenon construction. Member 'A' shows tenon, member 'B' is beam. Member 'C' shows mortise in beam 'B'

▲ Figure 3.6

Typical 18th century mortise-and-tenon wood framing

moment and provide stability. Some of the simpler mortise-and-tenon connections such as the one shown in Figure 3.5 illustrate the level of skill needed to achieve a structurally sound connection. These connections all typically involve the creation of a tenon, or flange, projecting from one piece and a tightly fitted cut out mortise in the second piece to perfectly receive the tenon. Often, an additional mechanical connection, stepped profile, or locking pin is used to further ensure that the connection is able to resist lateral forces as well as moment against bending. A minimum level of skill and workmanship was necessary to achieve successful mortise-and-tenon connections in building construction until the advent of the cheaply manufactured nail, in the early 1800s (see Figure 3.6). Milled lumber, in standardized sizes, furthered the modernization of frame construction.

There is some disagreement over the exact inventor of the balloon frame, though it came into use in the early 1800s, as sawmill production increased, with lumberjacks bringing a seemingly endless quantity of good-quality wood to market. Attribution has been given to George Washington Snow, by Sigfried Gideon, and perhaps more credibly to Augustine Deodat Taylor, a Chicago carpenter who was rapidly able to erect after the Great Fire of Chicago in 1871. Accounts of initial failures and sagging assemblies notwithstanding, Taylor used nails in combination with regularized 2" wide by 4" deep wood profiles in framed construction that could be erected quickly in response to great demand. Various patents and developments from the late 1700s through the early 1800s resulted in a fall in the price of the nail. Jesse Reed, for instance, invented a machine patented in 1807 for the production of up to 60,000 good-quality nails in one day.[1]

With the inexpensive nail replacing the tricky and time-consuming woodworking required of mortise-and-tenon construction, it did not take long for builders to recognize that closely spaced sawn wood could be used structurally without the heavy post-and-beam timbers. Though initially derided as thin and insubstantial, and therefore like a balloon, this method of construction took hold in North America. Framed only with small, but closely spaced, wood members, called studs, rafters, and joists, this construction method was economical, as it took perhaps as much as 40 percent less material than heavy timber construction. In addition, all connections, including finishes, could be made quickly using nails, by relatively unskilled labor, with a small number of workers carrying relatively small, light, studs and construction elements. This was in marked contrast to the heavy timber framing used in historic wood framing. Large numbers of workers, in addition to skilled carpenters to make the mortise-and-tenon joinery, are required for heavy timber framing, illustrated in Figure 3.6 and the photograph in Figure 3.7 from a barn raising for an enormous structure.

▶ Figure 3.7

Raising large two-storey balloon frames required many workers

Shipping of the raw materials was made possible by the new network of railroads which brought milled wood, nails, and the people they would house to the North American West throughout the 1800s. The construction of frontier towns expanded from the railroad depots, typically along a main street. The ease of construction and flexibility inherent in the light wood frame meant that newly arriving settlers could purchase the materials and erect a home, one-storey at first, in a very short time with minimal help, since all of the components were relatively small and light. Often, two-storey facades were erected in a single shot, even if the building to the rear could only be completed as a one-storey structure. When future prosperity permitted, a second floor could be added to back the false fronts with habitable spaces (see Figure 3.8).

Balloon framing

Balloon framing was the first step in creation of a kit of parts for residential and vernacular light frame construction. Derided as light, flimsy, and "balloon-like", in comparison with heavier and costlier alternatives, the speed of construction and efficiency of materials required for construction could not be overlooked.

▶ Figure 3.8

False Fronts, Niobrara, Knox, NE, 1880s

The primary components of a balloon frame, illustrated in Figure 3.9, consist of two-storey-high wood studs, sitting on a foundation wall, with diagonal bracing let into the vertical studs, and sheathing nailed to the outside, triangulating the assembly into a rigid, moment-resisting assembly. Floor framing joists were fastened to the vertical studs, and subflooring could be laid on top of the joists, stabilizing the floor assemblies in a similar fashion. Since the wood can be cut and reframed, this construction method provided significant flexibility for future expansion or modification. Tar paper and simple siding provided weather protection and rudimentary vapor control at the outside face of these structures (see Figure 3.10), until the advent of modern building paper and industrially produced insulation materials like fiberglass and mineral wool.

Wood frame construction was the fabric of the Victorian era, in part due to the ease with which cutting "gingerbread" decorative edgings and trim to residential construction could be produced. The backlash against this, in the later part of the 1800s, was the shingle style, also based in the same balloon frame construction method. Simplifying the decoration, returning perhaps to the early Americana made popular after the first U.S. centennial celebrations of 1875, houses both small and increasingly large adopted this volumetric expression of the building form. McKim, Mead & White's design for Isaac Bell's house in Newport, Rhode Island, from 1883, combined a return to simplicity, in the natural shingles, stained to a uniform monochromatic color. The small wood shingles, used on both vertical surfaces, the roof, and the many eaves and porches, also wrap around the cylindrical drum and cone-shaped shelters of this luxurious summer home for the wealthy cotton broker and investor. Nestled between the trees of the property, the wood siding, showing in Figure 3.11, is considerably more restrained than the Gilded Age resort mansions of the Vanderbilts and Astors.

Platform framing

Platform framing evolved as a variation on balloon framing. Balloon framing is still used in parts of North America, though smaller components can be used and there is no need for the more time-consuming construction of diagonal bracing with platform framing. It requires even less carpentry skills and bracing to achieve sound construction, and uses shorter framing members. The availability of wood being dictated by transportation limits also contributed to the development of platform framing. In platform framing, walls are assembled on an initial platform, then an upper platform is installed with subsequent wall framing tilted up, and so on. This means shorter, one-storey-high framing members are

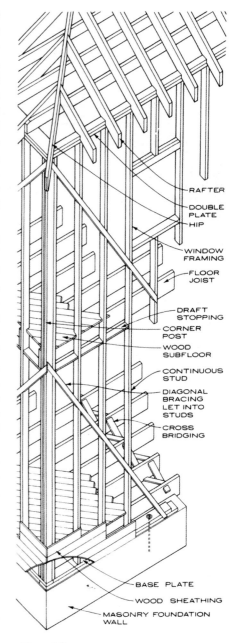

▲ Figure 3.9

Balloon framing terminology

▶ Figure 3.10

Clapboard siding. Walker Evans photograph, Shady Grove Baptist Church, Alabama, 1936

▶ Figure 3.11

Isaac Bell House, 70 Perry Street, Newport, RI, McKim, Mead & White, 1882–3, exterior view from southeast

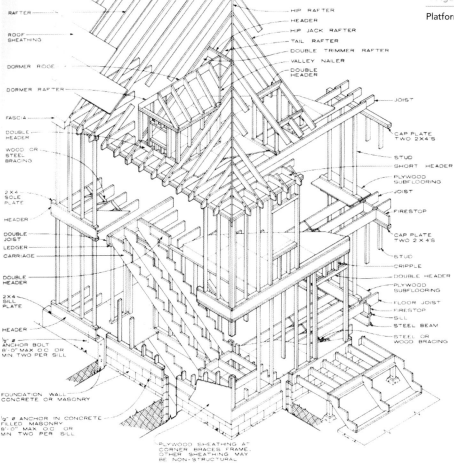

◀ Figure 3.12

Platform framing terminology

sufficient. Today, the most commonly used length is typically 8', short enough to be easily carried by one person, and requiring less bulky shipping, though longer lengths are also easily obtained, usually stocked in increments of 2' (see Figure 3.12). Shorter lumber can also be milled from newer growth wood.

While the virgin forests of North America have been reduced since the arrival of European settlers, more than one-third of North America is still covered by forest and woodland.[2] Though still readily available, responsible harvesting and renewable reforestation of harvested trees has started to address the need for long-term management of the raw materials for this industry in North America. Most timber produced in the United States today is harvested from private forests, both native and plantations. Concentrated in the eastern part of the continent, sustainable forest management is used to ensure continued production of timber, and to protect and provide ecosystems for wildlife and maintenance of water quality. Voluntary programs involving third-party certification include the American Tree Farm, the Sustainable Forestry Initiative (SFI), and the umbrella European organization Programme for the Endorsement of

Forest Certification (PEFC). A leading forest certification program, founded by environmental organizations in the early 1990s, the Forest Stewardship Council (FSC), implements even more rigorous standards and procedures for tracking chain-of-custody to certify that wood used in construction products was responsibly harvested. Wood products can be sold with the FSC label and used to meet LEED requirements and sustainability standards

Dimensional lumber commercially available to the construction industry was gradually standardized, starting in the early part of the twentieth century. By the 1950s, North America had adopted the two-by-four, or two-by-x as the most common dimensions for lumber. Available in two inch increments, typically 2 × 4, 2 × 6, 2 × 8, 2 × 10, and 2 × 12, these "nominal" dimensions today are given for green wood. When cut to these nominal dimensions, shrinkage of the green wood during the curing process reduces the actual dimension. Smaller pieces of wood from faster-growing plantations than the original virgin forests also contributed to the popularity of the platform frame wood stud construction method, which can be constructed with shorter studs than balloon framing, with lighter components that can be carried by fewer workers. The plantation-grown wood must still meet the strength grades established by manufacturing and accreditation programs, though typically softer wood than the original hardwood forests might have provided. Today, rule of thumb charts and recommendations have been developed based on common loading scenarios for standard stud sizes with regular spacing – 12" through 24" on center, most commonly – to identify height limits for a given assembly. Cross bracing can be added for additional strength, where needed.

The depth of the studs is largely determined by the need for the light framed wall to be able to resist lateral forces, perpendicular to the plane of construction. Stiffness is a concern when light framing is used behind less flexible materials like brick veneer, so sufficiently tight spacing between studs, and sizing of sufficiently deep framing members must be carefully executed to limit deflections. Sheathing material can be applied to both sides of the wood framing, forming a triangulated light structurally sound box.

The cavity created between studs becomes an opportunity to introduce insulating materials. Today, insulation is manufactured to fit between standard stud openings, friction fit with a vapor barrier (foil or paper) that can be installed on the warm side of the construction. In very cold climates, rigid insulation can also be added to supplement insulation between studs or in the case of metal studs, to minimize bridging of cold into a warm interior. Cold bridging is a concern, especially for metal studs that can transmit cold outdoor temperatures to the interior in spite of infill insulation between studs. The result can be excess moisture, if not outright condensation on the interior face of the wallboard. Traces of this can leave damp marks, cause collection of dust and dirt particles, and leave marks on the interior finish of screw locations, if not traces of full-height studs.

Efficiencies from engaging subcontractors to execute many similar houses with a limited set of options, but ability to purchase at larger economies of scale than individual homeowners, meant that developments by real estate owners proliferated in the years after World War II – by Levitt & Sons and countless others. Melding some of the garden city, utopian town aspirations of prewar planners

◀ Figure 3.13

Frank Lloyd Wright, Jacobs House, exterior view showing horizontality of wood siding

with profitable construction, developers such as the Levitts met the need of returning war veterans eager to marry their sweethearts and start a family in a new home. Often located just outside large cities, a variety of funding opportunities made the purchase of these homes financially viable with no deposit or down payment required, offering mortgages at good rates through veterans and other financing, and including washing machines and eventually even television sets built into the walls to qualify as part of the financed comforts.

Frank Lloyd Wright played on the abbreviation, "U.S.", in his "Usonian" houses, like the Jacobs House. These were modest homes constructed with light wood frame cavity walls, giving expression of the prairie house horizontality in relatively inexpensive vernacular wood construction. The exterior view in Figure 3.13 illustrates the horizontality created by the use of exposed siding boards at the Jacobs House. Like the earlier prairie houses, the Usonian houses shared a hearth-centered layout with an open plan, and some thought to solar orientation. The interior was a fluid extension, fully integrated with the exterior landscape. This optimized the small plots on which they could be built and enlarged the visual sense of the habitable space, though they were typically modest in footprint (see Figure 3.14). The detailing of the exterior wall incorporated

▲ Figure 3.14

Frank Lloyd Wright, Jacobs House, interior view showing continuation of horizontal expression at interior

▶ Figure 3.15

Frank Lloyd Wright, Jacobs House, axonometric showing wall construction. A – pine board. A low-grade wood was used. B – redwood batten. The darker redwood sets off the batten on the exterior. C – building paper. D – building paper. E – corner joint. In order to maintain continuous horizontal bands, wright used a miter joint here, rather than a batten as in the horizontal joint. This joint has a tendency to open up as the wood moves. F – base trim. Wright used the traditional projecting base to cover the wall-to-floor joint. G – concrete slab. Wright omitted the traditional footing in favor of a gravel trench.

human scale and touch, with the cantilevered bookshelf blades expressing the horizontal bands of the exterior boards at the interior face as well. The horizontal reveal on the wall exposed to the outside was used to support and bracket the shelf surface, creating floating horizontal lines within (see Figure 3.15).

Economies of scale from the war effort were applied to home construction in developments outside major cities, for instance Levittown, as built by Levitt Brothers. Large numbers of houses could be started and completed rapidly using efficiencies of crews producing first the foundations, then the framing, subsequently the sheathing, siding, and interior fittings in a systematic way (see Figures 3.16 and 3.17). New plots could be marked out and developed in subsequent years, with continuing improvements, additions, and the next year's new models, or layouts, from which the prospective homeowners could select.

Built for families ready to escape from the large city, into the new suburbs, Levittown was constructed at a time when personal cars became increasingly common and affordable, ready to navigate the new highways being developed, crisscrossing North America. Returning soldiers from the war were offered assistance with mortgages requiring little or no down

◀ Figure 3.16

Levittown construction

PART 1 Load-bearing construction

▲ Figure 3.17

Levittown street – newly completed Cape Cod houses

▲ Figure 3.18

Levittown aerial view – cookie cutter Cape Cod houses under construction

▲ Figure 3.19

Levittown aerial view – scale of successive adjacent developments

payment, and each successive year the developers could develop a new set of options for semi-customization of brand new houses to entice prospective buyers (see Figure 3.18). The population shifted into the suburbs, lured by the idyllic green family communities, good schools, and initially inexpensive land that often had recently been farmland. Though eventually there would develop a backlash against the isolation of an automobile-based society, the convenience and inexpensive, spacious homes, each with their own plot of land, made the houses attractive to many (see Figure 3.19).

◀ Figure 3.20

Gropius House 1939, Walter Gropius, 68 Baker Bridge Road, Lincoln, Middlesex, MA, exterior general view from southwest

▲ Figure 3.21

Gropius House 1939, Walter Gropius, 68 Baker Bridge Road, Lincoln, Middlesex, MA, view of west elevation

▲ Figure 3.22

Gropius House 1939, Walter Gropius, 68 Baker Bridge Road, Lincoln, Middlesex, MA, exterior second-floor west deck, view towards northwest

At the house built for his family when they emigrated to America in Lincoln, Massachusetts, Walter Gropius combined the hygienic, scientific housing theories of the 1930s and translated them in America into wood frame construction – though still tastefully sleek (see Figures 3.20, 3.21 and 3.22). Incorporating a masonry enclosure for the fireplace and chimney, the vertically oriented white boards siding the exterior of the wood frame lent sophistication and modern living to the suburban American landscape, using the same vernacular platform construction method.

Wood framed house

Our example house construction sequence starts with the foundation. As the base of the building, the foundation needs to anchor and spread loads from the house above onto soils below. Depending on site conditions, soil conditions, cost, and other criteria, the foundation can be poured concrete or concrete in combination with concrete masonry units. In this example from the northeastern part of North America, the excavation area for the foundation was wide enough to allow working clearance all the way to the foundation level typically required. In cold climates, the footings must be set deep enough to clear the frost line and prevent the entire superstructure from moving and eliminate heaving of the foundation with freeze/thaw cycles. Gravel is back filled around the footing, along with a perforated pipe for perimeter drainage all around the foundation. Filter fabric wrapped around punctured PVC pipe is commonly used to ensure that moisture or groundwater can be collected into the pipe. Filtration fabric prevents excess dirt and soil from being drawn into the pipe while still collecting water by setting it with a slight slope toward a low point or tank where a sump pump can be used to move the water in a controlled fashion into a municipal water system, or otherwise managed.

The foundation is typically covered with a parge coat of damp-proofing, as a minimum, if not full waterproofing in wet- or low-lying areas. If a basement is planned, rigid, waterproof insulation boards can be set outside the waterproofing of the exterior wall, to insulate the wall and the interior space from the cold ground outside, with condensation drained away from the heated interior face. Where no basement is required, filtration fabric in lieu of insulation can be used to protect the exterior face of the foundation wall. Threaded rods are grouted into either filled CMU foundation wall or poured-in-place concrete foundation wall (Figure 3.23), ready to anchor the starter sill for the light wood frame construction above.

The first-floor platform is then set on the perimeter foundations, establishing a plumb, level, and true plane on which to construct the first-floor framing. Unlike early heavy timber construction with mortise-and-tenon moment connections, modern light frame construction is simply nailed together, flat on the platform. The lead carpenter measures out the stud locations, typically 16" to 24" on center, onto sole plate and top plate studs, depending on loading conditions. The frame would then be nailed to these two wood members by the rest of the crew using 2" × 4" nominal stud dimensions, commonly: the frames create a ladder defining the exterior walls, with openings cut out and framed using short knee wall studs, rough sill plates to support windows at the bottom of the opening, and deeper headers (2" × 6" up to 2" × 10", usually nailed together in pairs to provide spanning stiffness) over window and door openings to brace the window/door and transfer vertical loads from above onto the studs at each side. After being nailed together flat, the frame is then tilted into a vertical position (see Figure 3.24). This way, an economical, small workforce can erect framed surfaces easily.

The framing tends to be very flexible until the sheathing is applied to triangulate the assembly, locking it into stability. Temporary cross-brace diagonal studs are therefore used to hold the frame in place until the remaining first-floor framing is completed to brace the open frame sufficiently, or until sheathing is installed (see Figures 3.25 and 3.26). The first-floor walls are framed in sequence, nailed

▲ Figure 3.23

Foundation ready for wood frame to be set

CHAPTER 3 Light frame construction

Wood framed house *continued*

▲ Figure 3.24

Light wood stud framing

▲ Figure 3.25

Light wood stud framing with temporary diagonal bracing

▲ Figure 3.26

Cross bracing for temporary shoring of wood framing

Wood framed house *continued*

to each other, with interior doors and openings framed in a similar fashion. The cage of framing thus forms a load-bearing support for the next level of platform construction, if desired (see Figure 3.27). The platform consists of spanning beams and joists, which are deep enough and rigid enough to bear the loads of the structure above as well as resist live loads from future inhabitants. A plywood deck is used to surface the top of the floor framing, forming a clean, level horizontal subfloor upon which second-floor framing can be set. Figure 3.28 shows sheathing applied to the lower level, then construction of the upper level platform. The second-floor framing is erected on top of the second-floor platform, as shown in Figure 3.29, and likewise braced until sheathing can be applied.

▲ Figure 3.27
Load-bearing first-floor framing for platform above

▲ Figure 3.28
Sheathing applied to first-floor framing

◀ Figure 3.29
Second-floor framing erecting on top of second-floor platform

CHAPTER 3 Light frame construction

Wood framed house *continued*

Figure 3.7 showed an enormous balloon frame wall, being tilted up as in a traditional barn raising, an event common in rural North America in the eighteenth and nineteenth centuries. At a barn raising, a collective community event bringing together sufficient hands to safely manipulate the framing in question, all able-bodied members of a town would gather to support one family's barn construction, with the understanding that the favor would be returned when the next family in the community needed the assistance. It becomes clear from looking at the sheer numbers required to lift up the heavy framing elements in the photograph from the erection of an extremely long, wide balloon frame assembly that the platform framing more commonly in use throughout North America today is significantly lighter. A surprisingly small workforce is able to execute nearly the entire construction of light frame construction. Specialized skills are typically only required for sitework, plumbing, electrical, and heating or cooling, besides general and finish carpentry.

Once the final platform is framed, ceiling joists to support upper level ceilings, and then rafters to support a sloped roof are shown partially installed in Figure 3.30. Many shapes can be configured for roof profiles with studs, and prefabricated trusses or engineered wood profiles and beams, like those shown in Figures 3.31 and 3.32, can be used in conjunction with 2" × 8" and deeper wood studs, small steel columns, or beams.

◀ Figure 3.30

Ceilings and roof rafters are framed for interior finish and roof support

▲ Figure 3.31

Prefabricated engineered wood

▲ Figure 3.32

Prefabricated engineered wood

Wood framed house *continued*

▲ Figure 3.33

Interior of stud framed construction looking at exterior sheathing. Insulation can be friction fit between studs

▲ Figure 3.34

Wood framed house with expansion in progress. Asphalt impregnated black building paper is lapped from foundation up to roof eaves. Clapboard siding will be installed on building paper to extend the original water-shedding siding

A modular system for sheathing based on the 4′ × 8′ dimension is prevalent for modern wood frame construction in the U.S. A system for nailing on even modules ensures that even waste can be reused if it matches the standard stud support module. At corners, configurations can be achieved within wood stud detailing to provide nailing surfaces for attaching exterior sheathing and interior finish materials, alike. At the interior, this is most commonly gypsum wallboard. See Figure 3.33 for a view from the interior of a stud framed structure, looking back at the interior face of the sheathing.

There are many types of sheathing available, whether plywood-based, gypsum, or fiberboard (oriented strands of wood, thus Oriented Strand Board or OSB). Each has different properties, strengths, and weaknesses, and should be selected carefully based on different exposures and uses. To supplement the structural sheathing, building wrap and asphalt-impregnated building paper are used to improve the moisture resistance of the sheathing. Layers of the air-permeable material are wrapped around the building, tacked into place with staples. Each layer is rolled out, applied in rows from foundation upwards to ensure positive drainage of any moisture. The building paper and house- or building wrap are not typically suitable for extended exposure to UV/sunlight, but improve the performance of the skin and increase protection against water infiltration by providing a backup layer immediately below the finish siding. Figure 3.34 shows an example of a wood frame house expansion and renovation with lapped black, asphalt impregnated building paper over a brick foundation. Roofing has been installed, and siding will be added next. Note how the building paper is tucked under the clapboard siding of the existing house, to the right, to maintain continuity of moisture protection.

Installation of windows and doors is the next step in closing the exterior envelope of the house. Figure 3.35 shows a house under construction with roof shingles lapped to shed water; building paper lapped over the roofing; windows installed with flashing to drain any moisture from the opening out and over the building paper; and copper head flashing above each window to ensure water will not penetrate to the interior through the top of the window. During construction, care is taken to ensure that each subsequent layer laps over the previous one, to ensure that rainwater and moisture is drained away from the inner surface of the wall (see Figure 3.36).

◀ Figure 3.35

Installation of flashing around windows occurs prior to installation of finish siding to ensure positive flow of moisture from leaks at window openings over building wrap

Wood framed house *continued*

◄ Figure 3.36

Installation of shingles from foundation upward to ensure overlap of finish siding and proper drainage of rainwater

▼ Figure 3.37

Architectural trim and detailing to further protect window openings from water infiltration using projecting trim and special treatment of shingles over window head

Finish siding is applied for weather protection of the stud wall assembly. Water should be allowed to cascade down the surface of the final finish, whatever the material or configuration. In a completed house with cedar shake siding (Figure 3.37), recessed windows at the ground floor and an unusual eyebrow projection of the siding, or more conventional wood trim provides further rain protection, architecturally. If wood is used to finish the outer face, it may require some protection, itself, in the form of paint. Where left exposed to elements, the finish material may be subject to water in the form of rain, snow, ice, and hail, in a cold climate. The exposed finish siding will also be exposed to wind and temperature swings from the climate of the region, and any microclimate effects of topography or exposure. Depending on sun, shading, vegetation, or other surrounding buildings, the finish may also be exposed to other environmental variations. Note how in this cedar shake house the unpainted shingles have weathered more quickly, turning grey at the south face, while the western face and shaded surfaces, shielded from light, have not weathered as quickly as at the south, with its increased sun exposure, as shown in Figure 3.38, for south and east exposures.

At the interior face, after the trades have completed installation of piping, conduit, and any heating/ventilation/air conditioning ductwork, insulation manufactured for wood frame construction shipped in soft batts can be friction fit in between standard wood studs and is available for most standard stud spacings, with either stiff paper or foil at the interior face. The batt insulation can then be taped with paper or foil tape, to seal joints between rolls and between studs, creating continuity of vapor barrier impermeable to air and moisture alike. An alternate is to wrap thin polyethylene plastic sheeting against the interior face of the studs and insulation, prior to installation of the interior finish sheathing, typically gypsum wallboard, commonly called sheetrock or drywall. Spackling and taping of the standard, modularized gypsum wallboard after being screwed into the studs, results in a clean, smooth surface ready to receive paint or other interior finish treatments. Finish molding hides the gaps between windows and interior gypsum wallboard at vertical surfaces, ceilings, and flooring, where the gaps are also important to permit expansion and contraction with humidity changes in the exterior and interior environment.

Wood framed house *continued*

▲ Figure 3.38
Uneven weathering visible in new cedar shingles – evidence of difference in sunlight exposure, prevailing wind and rain

▲ Figure 3.39

Stucco – construction view of lath being installed over sheathing

Options for the final exterior finish in a wood frame assembly can vary from the most traditional North American examples as shingles, shakes, board-and-batten, or clapboard siding; through stucco on lath; veneer masonry; or almost any material that is environmentally stable with exposure to the elements and easily manipulated into desired configurations. Notable developments in recent decades which should be considered carefully with respect to thermal and moisture performance, as well as life-cycle and environmental impact from manufacture and disposal, are composite sidings, including asphalt, asbestos, fiber cement, and other materials; aluminum and vinyl siding, reproducing the appearance of wood without requiring regular painting; and insulated foam with artificial stucco surfacing, or Exterior Insulation and Finish Systems (EIFS). In Figure 3.39, we see an example of a building that has been enclosed by sheathing and fully roofed, ready to receive galvanized mesh, which would support the final Portland cement stucco finish shown in Figure 3.40.

At Seaside, Florida, Duany/Plater-Zyberk used an urban zoning code to encourage a certain consistency of aesthetics and finishes of largely wood frame houses and community buildings. Zoned variation was used to avoid the cookie cutter monotony of typical developer housing, encouraging density and applying concepts of town planning. The development used the configuration of conventional wood frame house construction for the purpose of fostering interaction between neighbors and established a very desirable, pristine community with echoes of Southern grandeur as well as New England clapboard simplicity. The division of property into small plots encouraged interaction and achieved elevated property values as a result. Exploiting nostalgia for the effect of such

▲ Figure 3.40

Stucco finish – completed exterior view

small town living, the development created a town with an image of exclusivity and charm to a very successful, profitable effect (see Figure 3.41).

Plywood was used as exterior finish in Herzog & de Meuron's projects such as the Plywood House, Bottmingen, Switzerland, completed 1985 (see Figures 3.42 and 3.43). One side of the otherwise rectangular house addition is bent, in plan, to acknowledge a mature, existing Paulownia tree. Using stained plywood as the exterior siding and a light wood veneered plywood for the interior finish, the addition is articulated like the puppet theatre it was designed to house.

A notable variation on wood frame construction using light, small elements combined to achieve many effects is the chapel in the woods at Eureka Springs, Arkansas, 1978–1980, designed by Fay Jones. Called Thorncrown, this nondenominational chapel elevated mundane wood frame construction in a sensitive use of material local to the region. Transparent infill glass encloses the interior, which emphasized verticality like the surrounding trees. All of the structure is exposed, including bracing used to triangulate and stiffen the entire cross section of the church into decorative geometry drawing the eye upwards (see Figures 3.44, 3.45 and 3.46).

A variation of plywood manufactured from wood layers alternating with glue layers is another sub-category: engineered wood products. These materials are made from thinner, smaller wood engineered and sandwiched with multiple layers of bonding glue into larger, longer, sometimes thicker, and usually stronger profiles and assemblies. Laminated veneer lumber or glue-laminated wood can achieve greater consistency of strength than natural wood, which is subject to weakness if wood growth was too rapid, as moisture content swells or

▲ Figure 3.41

Seaside, FL, Duany/Plater-Zyberk

▲ Figure 3.42

Herzog & de Meuron Plywood House, Bottmingen, Switzerland

▲ Figure 3.43

Herzog & de Meuron Plywood House, Bottmingen, Switzerland

▶ Figure 3.44

Fay Jones, Thorncrown Chapel, view from exterior

CHAPTER 3 Light frame construction

▲ Figure 3.45

Fay Jones, Thorncrown Chapel, night – interior view

▲ Figure 3.46

Fay Jones, Thorncrown Chapel, night view from exterior

shrinks, and behavior under load occurs. These laminated wood products also tend to achieve better moisture and absorption performance than wood alone. The fabrication of these products requires more added labor to raw wood than simple milling and drying of lumber, with attendant waste and byproduct, but also offers a method to reuse waste from first-round milling and create viable construction materials with calculated, reliable strength. Glue-laminated wood is marketed in proprietary sizes and construction components, whether for bridging greater spans in residential construction in lieu of joists and beams, or even replacing steel, in some cases. Glue-laminated wood beams can be used for larger spans to interesting effect, taking otherwise waste materials or smaller, newer growth wood to reliable effect.

Outstanding examples showcasing the use of engineered wood products include the speed skating ice rink, Vikingskipet, or The Viking Ship, at Hamar, designed by Niels Torp Arkitekter and Biong & Biong (now Biong Arkitekter), constructed in 1992 for the 1994 Winter Olympics centered in nearby Lillehammer. This building incorporated deep glue-laminated beams to shelter the skating oval in a roof with a profile reminiscent of an upturned Viking boat (see Figure 3.47). The form of the roof is shaped by a longitudinal metal roof broken by stepped slits, reminiscent of wood planks forming the keel and hull of a boat (Figure 3.48),

▲ Figure 3.47

Niels Torp/Biong & Biong, Vikingskipet, Olympic Speed Skating Rink, Hamar, Norway

▶ Figure 3.48

Niels Torp/Biong & Biong, Vikingskipet, Olympic Speed Skating Rink, Hamar, Norway

▲ Figure 3.49

Niels Torp/Biong & Biong, Vikingskipet, Olympic Speed Skating Rink, Hamar, Norway

◀ Figure 3.50

Studio Gang Architects, Starlight Theater, exterior view

while the curving transverse trusses stiffen and support the roofing like the ribs of a Viking ship. The graceful shallow curved ship hull encloses the rink and its support spaces while echoing the national heritage of Norway (see Figure 3.49).

The Bengt Sjostrom Starlight Theater, designed by Studio Gang Architects for Rock Valley College, Rockford, Il, also used glue-laminated roof construction (see Figure 3.50). At this college campus triangular roof panels, set over an open-air summer theater seating bowl, open like petals of a star-shaped flower.

▶ Figure 3.51

Studio Gang Architects, Starlight Theater, construction view showing underside of triangular panel with glue-laminated beams

The glue-laminated beams stiffen each facet of the roof, sheltering the stage and seating below so that performances can proceed, rain or shine. The roof panels are surrounded by reinforced concrete walls whose reinforcing bars were configured to allow round openings lit like planets and stars. When the weather permits, the star-shaped roof opens to the sky during theater performances (see Figures 3.51, 3.52 and 3.53).

Light frame construction with wood studs and wood-based sheathing is used in vernacular, residential, and generally smaller-scale construction because of a number of factors. Not requiring highly skilled knowledge, generally speaking, the basic wall construction method is commonly achievable for the informal

◀ Figure 3.52
Studio Gang Architects, Starlight Theater, interior view of roof panels opening

▼ Figure 3.53
Studio Gang Architects, Starlight Theater, exterior view

▲ Figure 3.54

Metal stud framing

do-it-yourself frontiersman and the suburban weekend homeowner, alike. Though commonly used in single-family dwelling construction where the materials are plentiful and relatively low cost, most jurisdictions have adopted building codes that do not permit combustible framing materials to be used for multi-family housing or larger commercial buildings which might have higher risks of flame or fire spread. Thus the development of metal studs for comparable, equivalent lightweight construction purposes with lightweight materials that can be easily transported and manipulated has become ubiquitous in interior partition construction where combustibility is a concern. These profiles are generally available in similar standardized shapes. In North America, these sizes are often based on wood stud dimensions.

The evolution of lightweight steel studs for commercial, retail, and other purposes has in part been due to the noncombustible nature of the material, in part due to the low cost and high flexibility of the framing. In Figure 3.54, we see that light gauge steel studs can achieve very straight, plumb construction with many small, lightweight elements. They are able to span even great distances with more rigid simple channel shapes, called C-studs, but are available in a wide variety of proprietary profiles that are relatively easily transported and installed. These studs are often used in conjunction with gypsum and sheathing products in a myriad of ways, from light industrial buildings and "big box" retail, sometimes with corrugated metal siding and minimal insulation.

Unless detailed carefully, some of the hazards associated with metal stud exterior wall construction include thermal performance and cold bridging concerns. Figure 2.11 in Chapter 2 shows an assembly with metal stud that has the potential to be highly problematic. In this example wall section detail insulation within the cavity must raise the dew point sufficiently or the metal may carry cold into the interior of the building, which could cause condensation to occur on the interior face of the metal studs, against the interior wallboard shown. Combining such a light frame strategy with outboard insulating in a cavity wall configuration would help resolve concerns regarding condensation and thermal bridging. Light gauge stud construction has infinite uses for interior dividing partition construction and fit-out of multi-family housing, retail spaces, and offices, especially where there is a concern to limit fire spread between units since the material is fundamentally not combustible, though, like all steel, it can lose strength when exposed to high heat in a fire.

By sizing sufficiently heavy gauges of the stud material thickness and sufficiently closely spacing them, it can be possible to pair light frame construction with the rather unforgiving masonry, which cannot accommodate large flexure deflections from lateral forces, even with conventional reinforcing mesh. The placement of insulation should be carefully detailed with vapor barrier on the warm side of the insulation and flashing to carry condensation to the exterior. This strategy can be economical yet provide advantages of the cavity wall, which splits the building skin into a two-layer assembly with a cushion of air, insulation, and vapor barrier, between the outer veneer and the fully insulated inner structural wall.

Part 2 | Non-load-bearing construction

Chapter 4
Frame and infill construction

The development of wall construction types from ancient examples through the early twentieth century evolved as methods of assembling materials into increasingly sophisticated systems gave rise to control of such physical properties of nature as heat loss, vapor transmission, and moisture management. The rapid leaps of material production of the industrial revolution, which encompassed methods of harnessing naturally found materials like wood, clay, and iron, took an additional leap in the course of the twentieth century as postwar factories turned away from war production. New materials and new ways of producing old materials became a rich new source of construction materials, with accompanying changes and advances in the ways of addressing the same physical properties of the natural world.

As we saw in the heavy, load-bearing masonry systems, traditional materials can be used to achieve high spans whether with brick arches or stone buttresses, such as in the Sainte Chapelle, Paris, twelfth century (Figure 4.1), an example using stained glass as infill to masonry arches. Like Notre Dame cathedral, the stained glass windows at the small chapel built by Louis IX as a holy chapel, or Sainte-Chapelle, attached to the palace that stood beside it, rise higher and lighter with thinner stone piers, thanks to heavy outboard buttresses

▶ Figure 4.1

Sainte Chapelle interior

◀ Figure 4.2

Sainte Chapelle exterior buttress

▼ Figure 4.3

Sainte Chapelle interior

(Figure 4.2). The thirteenth-century stained glass infill between stone columns (Figure 4.3) allows light in through the tall Gothic arches, creating a sense of weightlessness and verticality to the stone construction.

The introduction of new materials for architectural use meant the capacity for much greater tensile strength, reaching equivalent heights if not higher spans with lighter elements. The shift from heavy, load-bearing methods of construction towards lighter, mechanized assembly methods by the nineteenth and twentieth centuries was also reflected in architectural theory and fashions, alongside the fascination with modern, economical, and efficiently functional construction.

Timber precursors to frame and infill construction were previously introduced in our light frame chapter (see Figures 4.4 and 4.5). Ancient examples of timber framing from China and East Asia required skilled carpentry to achieve structurally sound mortise-and-tenon structural moment connections in wood frame buildings. The areas between framing members were filled with lighter, more fragile enclosure material, in some cases elaborately carved infill panels. The following examples also illustrate various types of infill systems within structurally self-supporting framing:

▶ Figure 4.4

Chinese mortise-and-tenon wood framing – terminology and proportions associated with the Chinese equivalent to an "order"

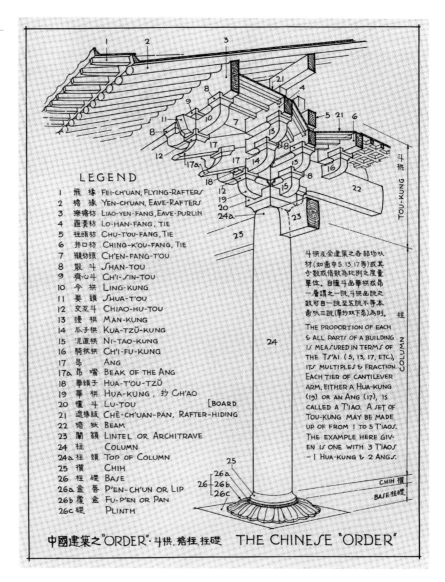

- A famous Japanese example, the Katsura Imperial Villa, outside Kyoto, Japan, sixteenth century (Figures 4.6 and 4.7), further extended this frame and infill tradition by incorporating the use of the modular "tatami" padded straw seating/flooring, establishing a unit used throughout the planning in an asymmetric arrangement.
- Western examples from medieval Europe, both Tudor and from other time periods (Figure 4.8) used an infill material that incorporated woven strips of flexible strong reeds, covered with a straw, clay, and/or dung mixture, also known as "wattle and daub" (see Figure 4.9).
- Norwegian stave churches, another post-and-beam example, used heavy corner posts and wood infill panels, and are clearly an example from a cold

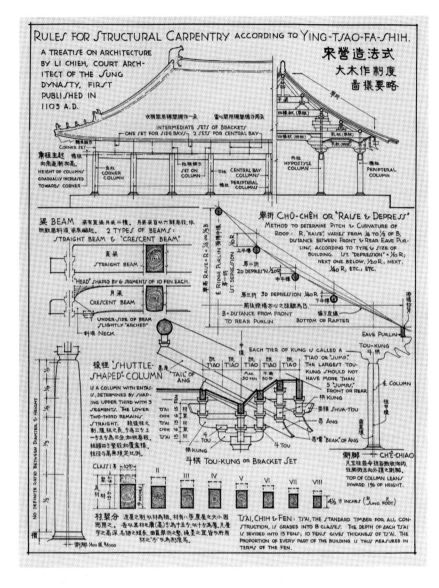

Figure 4.5

Chinese mortise-and-tenon wood framing – from a treatise on architecture dated 1103 AD

Northern climate typically covered with multiple, steeply pitched roof levels to shed snow (see Figure 4.10).

During the industrial awakening of the nineteenth century, James Paxton's scheme for the Crystal Palace in Hyde Park, England, was erected in record time in 1851 (see Figure 4.11). This project, which exploited the speed and efficiency of large-scale mass production, was executed in record time using modular glass and iron components to create an enormous structure for housing the Great Exhibition of 1851. Using knowledge of glass infill in large expanses from his experience as a gardener at Chatsworth House, whose greenhouse was a

▲ Figure 4.6

Katsura Imperial Villa, exterior

▲ Figure 4.7

Katsura Imperial Villa, interior

▲ Figure 4.8

Timber frame house, Becher

▲ Figure 4.9

Wattle and daub infill – photograph from repair

◀ Figure 4.10

Stave Church, Norway

precursor of the Crystal Palace, Paxton's design was large and tall enough to encompass full-size living trees. Construction of the soaring exhibit hall was constructed with a crew of workers using a kit of parts (see Figure 4.12).

Cast iron framing with large glass infill panels was brought to the construction of warehouse and printing houses for their fire-resistant capabilities in New York City's Soho for example where we see many remaining instances of these

▲ Figure 4.11

Crystal Palace

early department store and storage buildings (see Figure 4.13). The large areas of glazing in proportion to wall area were beneficial in providing daylight using a generally noncombustible construction type. These early "skyscrapers" as they were called at the time were functional precursors to the skyscrapers and tall buildings that evolved thanks to the developments in structural steel and reinforced concrete structural framing as well as the development of the elevator and improvements in indoor plumbing. The building at 55 White Street, constructed by Daniel D. Badger, used one of the many variants available through his Architectural Ironworks (see Figure 4.14). Double-storey columns expressed on the facades emphasized the verticality of these commercial buildings, where maximizing daylight minimized the need for hazardous lamps that might cause a fire, and resulting loss of goods stored within.

These precursors to the modern frame with infill method of construction used relatively large pieces of glass as infill in the same way that modern architects working with infill glazing and other materials did after World War II. The infinite Cartesian space envisioned by the later work of Ludwig Mies van der Rohe, for example, defined a spatial understanding with a frame structure where the structural grid was brought to the exterior surface of the building envelope. Diagrammatically, this relationship of the infill to frame was either expressed by projecting the edge of the floor beyond the remaining enclosure, like in the

◀ Figure 4.12

Crystal Palace, glazing

Farnsworth House of 1951 (Figure 4.15), or flush as seen in the later city center projects. In this structural frame with non-structural infill relationship, a support structure, or frame, takes the dead loads and lateral loads imposed by the infill cladding system. The infill itself requires support for its weight, and receives lateral wind loads which it transfers back to the main support frame.

The infill system must be designed to anticipate and accommodate differential movements and tolerances from manufacturing of the infill material as well as installation within the frame and live load movements of the frame structure. In addition, the thermal expansion properties of both the frame and the infill materials we studied in the previous chapter must be accommodated. To prevent the designed joint from appearing to shrink at some areas to a hairline joint (where all of the tolerance and expansion capacities are used), it is ideal to enlarge the joint so that the eye perceives a continuous joint at all times.

> JOINT SIZING RULE #1: for the best visual appearance, the actual joint should be twice the required actual joint dimension. See joint sizing box.

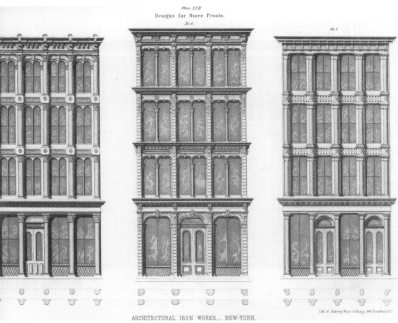

▲ Figure 4.13

Cast iron storefront, Daniel D. Badger

▲ Figure 4.14

Cast iron storefront – 55 White Street, Daniel D. Badger

▲ Figure 4.15

Edith Farnsworth House, 14520 River Road, Plano vicinity, Kendall, Il, Ludwig Mies van der Rohe, 1946–1951

◀ Figure 4.16

Edith Farnsworth House, 14520 River Road, Plano vicinity, Kendall, Il, Ludwig Mies van der Rohe, 1946–1951. South elevation, detail of entrance looking from the east

An infill system can be any one material or any assemblage of materials, including any of the systems previously studied – monolithic bearing wall, cavity wall, or light frame. Insulation values of the infill system can vary but its effectiveness is often negated by the fact that most exposed slab edge can become a cold bridge. Potential problems with exposed building frames can be best illustrated by such example projects as Mies's Farnsworth House, in particular, and countless exposed slab high-rise buildings, in general.

At the Farnsworth House, the exposed structural grid carries cold into the interior faces of the steel, bringing the extreme exterior temperature towards the interior (see Figure 4.16). These columns are exposed for their full height and are liable to cause condensation where the cold, impervious metal meets the warm and likely more humidified interior air. As we saw in the cavity wall chapter, when the thermal gradient intersects the dew point condensation will be drawn out of the air and form droplets of water. This naturally is far from

▶ Figure 4.17

Edith Farnsworth House, 14520 River Road, Plano vicinity, Kendall, Il, Ludwig Mies van der Rohe, 1946–1951. North Elevation showing framing of house, floating above ground plane

desirable against steel which is especially vulnerable to corrosion with long-term moisture exposure. In the exterior photographs, such as Figure 4.17, we can also see that the steel edge beam is likewise a vulnerable component. Since it is not insulated, yet connects from the interior to the exterior, the steel becomes a reverse radiator, carrying the heat in winter immediately to the outdoors and radiating heat loss to the outside environment. The solution to the condensation problem would normally incorporate insulation and a proper vapor barrier at the warm interior face of the insulation, for a cold climate, wrapping the entire indoor environment with a warm blanket and condensation barrier forming a continuous air seal to the exterior. However, it is unlikely to meet with any preservationist principles since it would be nearly impossible to maintain the architect's original profiles. For this project, which we now consider a landmark of modern architecture, regular replacement of damaged components as well as raising the interior temperature in order to offset the condensation and push the dew point outboard of the interior space of the dwelling may be the most respectful solution, though this likely means expending a great deal of heating energy in the process, the house having been constructed at a time when the energy costs of today were unimagined.

Whatever the infill material used to fill an exposed framing system, the material must be designed to accommodate global structural movements as well as any expansion of the infill material itself. Though we generally consider a brick-and-mortar infill to be solid and durable, it still responds to thermal swings over daily and yearly cycles by expanding and contracting according to its inherent thermal coefficient of expansion. The natural properties of metal dictate an even higher thermal expansion coefficient, so exposed slab projects such as the apartment building constructed by Jean Prouvé in Paris (Figure 4.18), in the Fifth Arrondissement, must include a slip joint to permit expansion of the infill material without construction. Deflection at head of the infill must also accommodate live load and any other building frame movements. These include creep, from long-term curing of concrete frame; global building movements from wind or other lateral forces due to building geometry; live loads that might cause downward movement from the structure above; as well as seismic loads in areas of the world where this is required. Seismic design for masonry typically also requires restraint

▲ Figure 4.18

Jean Prouvé, Apartment building, Paris, France

▲ Figure 4.19

Aqua Tower, Studio Gang Architects, Chicago, Il, 2010, exterior view

of some kind at the head, such as staggered steel angles, to keep the infill wall from tipping over while still accommodating structural deflection from above.

As with monolithic masonry used as infill, all the principles of self-support and all concepts of how the system is insulated and how water is flashed out of the system to the exterior still apply. Examples of exposed concrete slab construction with infill non-load-bearing, self-supporting infill are widely available in multistorey commercial and residential projects alike. Exposed slab edges are still vulnerable to freeze/thaw deterioration. Even if the infill properly drains moisture, the weep moisture can in fact increase the vulnerability of the concrete frame and slabs since small amounts of water flow over them, leaching into cracks of the porous concrete where the moisture may freeze, expand, and spall the concrete, rendering the reinforcing steel vulnerable to deterioration as well.

Figure 4.19 shows the Aqua Tower by Studio Gang Architects, a mixed-use project in Chicago, Il, completed in 2010, where the slab edge was manipulated to create a distinctive, undulating appearance. The cantilevered concrete slabs project from the face of the exterior enclosure line of the tower, which is defined by floor-to-ceiling glass captured in infill framing. Extending the slab in curvilinear edge that changes incrementally in curvature and profile gradually up the building, at each floor, the profile created is that of a sinuous, rippled surface emerging from the glazing. The balconies created by the projecting horizontal floor slabs (Figure 4.20) allow the zone between the glass and the implied surface to be inhabited by the occupants of the apartments and hotel rooms, and create sun shade for some portions of the facade, providing some offset to the heat loss through the balcony slabs.

▶ Figure 4.20

Aqua Tower, Studio Gang Architects, Chicago, Il, 2010, building section showing infill glazing between projecting slabs

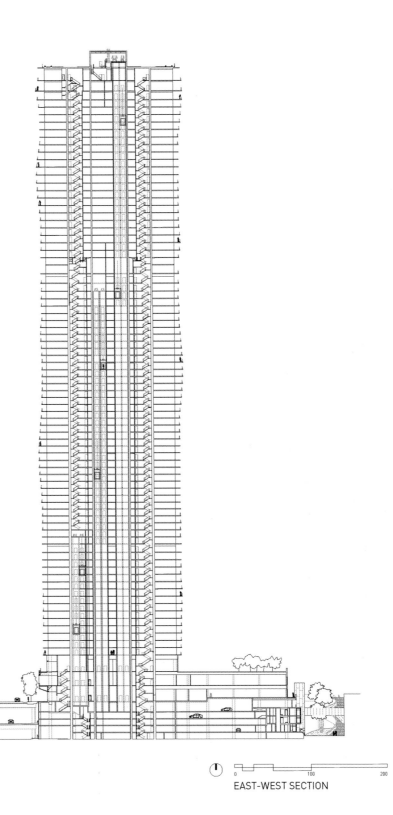

EAST-WEST SECTION

CHAPTER 4 Frame and infill construction

Glass

Commercial production of glass for architecture evolved from early hand-blown window glass through drawn methods in the early twentieth century, before the invention of industrialized methods that allowed the material to become economically cost effective for the masses. Hand blown into circular shapes (see Figures 4.21 and 4.22) that were spun out into flat planes roughly 5–6 feet in diameter, at most, then cut into smaller and smaller shapes to maximize the resulting flat pieces, crown glass was a precious material for centuries. From this process were developed the blown and drawn method of producing large cylinders of glass in factories, which were cut open while still semi-molten and flattened to produce larger and larger sheets of glass. Plate glass, which required time-consuming grinding and polishing of cast slabs, provided flat, optically clear glass, but at a great cost. Cylinder and flat drawn glass, introduced in the early twentieth century, were still

▲ Figure 4.21

Diderot *Encyclopedie* – Glaces Soufles, plate 36

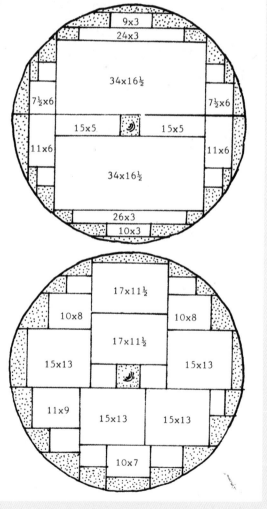

▲ Figure 4.22

W. Cooper Crown Glass

Glass continued

in wide use until the float glass process was invented and quickly became adopted worldwide.

The production of glass was revolutionized in the 1960s by Alistair Pilkington's industrialized float glass process, so called because the flat glass is made quickly and easily by floating molten glass over a bath of molten tin. Commonly used architectural float glass is often called Soda Lime Glass, because of its constituents: sand 75 percent, soda 13 percent, lime, and cullet (recycled crushed glass starter). This is melted and floated out onto a bath of molten tin. The speed at which it is spread dictates thickness, at standard increments (metric sizes used in industry, but translate to ⅛", 3/16", ¼", ⅜", ½", and ¾", typically). As the glass moves off the line, it is cut into manageable lengths and cooled slowly so that stresses do not build up unevenly in an annealing oven. The result of this production is therefore known as Annealed Float Glass.

Loads on glass

Lateral loads on glass are primarily wind gusts which put pressure and suction forces on the face of a building. In addition to these more commonly considered loads the sun can put thermal load on glass through uneven shading. Imagine a very cold morning, when the sun comes out and heats the surface of a building quickly. It is possible for a lite of glass to be subject to a significant temperature differential. In these easily overlooked situations additional treatment to strengthen the glass may be necessary.

Many issues in the selection of glass should be considered, starting with the desired appearance and use. Hard selection criteria must be based on the required size and strength, balanced against methods of fabrication and manufacture. Glass is manufactured by a limited number of factories but sold through a fabrication and distribution system where added work can greatly enhance but also limit the options for designing with architectural glass.

Strength

Annealed glass is taken directly off the float line and used widely in architectural applications. However, it breaks into sharp shards, which can be quite dangerous (see Figure 4.23). Heating the glass again under controlled conditions, then cooling quickly, causes the outer surfaces to go into compression while the interior glass slowly solidifies into a state of tension. The glass achieves a balance of higher compression at the outer surfaces and correspondingly higher internal tension forces internally. When the surface of this "fully tempered" glass receives an impact it is able to resist greater loads – typically four times stronger than annealed glass of the same thickness – because the surface is already under compression. When broken, the tempered glass shatters into "dice" roughly the same dimension on each side as the thickness of the glass (see Figure 4.24).

▲ Figure 4.23

Breakage pattern of annealed glass

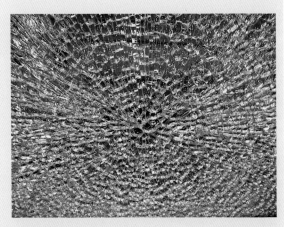

▲ Figure 4.24

Breakage pattern of tempered glass

These fractured bits of glass are less dangerous than the blade-like shards of broken annealed glass, therefore the glass is categorized as safety glazing (see Figure 4.25). An intermediate, called "heat treated" glass, is not reheated to the same temperatures as fully tempered glass and is therefore not able to produce safety glazing, but achieves strength on the order of approximately two times regular annealed glass, and has correspondingly less distortion when compared to fully tempered glass.

Assembly

Glass can be used in single, or "monolithic", thickness. It can also be sandwiched together with an interlayer filling

Glass *continued*

▲ Figure 4.25

Tempered glass breaks into small dice

▲ Figure 4.26

Strength of tempered glass

▲ Figure 4.27

Laminated glass retains broken annealed glass in a safe manner

that bonds the two lites together, using a PVB most commonly (polyvinyl butyral – soft strong laminate taken from automotive industry), resins (liquid cured by UV light exposure, therefore not safety protection, or two-part resins that may achieve safety ratings), or other material, usually bonded using heat and pressure. Interlayer film laminates typically fall into the safety glass categorization of most model codes since broken glass, regardless of heat treatment of individual lites, remains adhered to the laminate when broken and protects building occupants by remaining in the glazed opening in a relatively safe manner and providing some level of ongoing protection even if broken (see Figure 4.26). They replace the earlier wired glass, which is subject to additional issues from thermal expansion of the embedded steel mesh, and are better for retaining glass in a safe manner when broken (see Figure 4.27). This is particularly useful in overhead conditions where safety is always a concern.

Glass can also be assembled into insulating glass units (IGU), most commonly two lites sealed to a perimeter air spacer frame with two lines of sealant (primary seal made of soft butyl, secondary seal made of silicone sealant, dessicant in spacer). IGUs can be a very effective insulator because of the trapped air in the cavity, which is not wide enough to induce air currents. The additional surfaces also provide opportunities for adding coatings, patterns, or other treatment. Triple glazing is less commonly used but increasingly available, though multiple coatings should be analyzed carefully to prevent accidental visual effects (moiré interference patterns). Drawbacks of insulating glass units include the fact that the framed edges are not intended to be exposed to view, and can be unsightly. Size

Glass continued

▲ Figure 4.28

Wired glass – historically used to provide some protection in case of glass breakage

and weight limits should also be checked, since the fabrication limits and weight on the building can be a factor – it can be very difficult to accommodate the additional thickness in retrofit situations, though the thermal performance improvement to heat loss through the building envelope are considerable.

Treatment

A variety of treatments on glass are available affecting both the appearance and performance of glass, though some are not permitted for use on the exterior because of decay from UV exposure or uneven thermal loading on the glass, though it can be treated to better resist thermal differentials (at a slight cost). Fully tempered glass is approximately four times the strength of annealed glass (see Figure 4.28).

Textured – patterns can be rolled into hot glass, which allows light transmission but obscures view through the glass while simultaneously applying a pattern (dots, textures, leaves, etc.). This is done at a factory on a large scale, but usually has a limited size range and is not typically suitable for the exterior of buildings since the texturization reduces the thickness of the glass and its ability to resist lateral loads.

Ceramic frit – pigment is mixed with ground glass and painted or silkscreened onto glass, then heated to fuse

▲ Figure 4.29

Herzog & de Meuron, Ricola Factory, patterned translucent glazing at canopy and facade

the glass/pigment onto the surface of the glass. This should bond very well, but requires heat treatment of the glass to resist uneven absorption of solar radiation. This is often used at spandrels (100 percent coverage, also called "flood coat") and skylights (can be silkscreened in patterns to allow some sunlight in but provide partial shading effect). An interesting example of a custom pattern on translucent glazing material is the Ricola-Europe SA factory and storage building in Mulhouse-Brunstatt, France, 1996, designed by Herzog & de Meuron. The silkscreen printed pattern of a plant motif is repeated across the vertical cladding panels as well as the cantilevered canopy to filter the daylight and shelter the loading and transport on the long sides of the factory (see Figure 4.29).

Coated glass – thin metallic films are applied to the glass surface using a vacuum deposition process, for high performance coatings; or the older pyrolitic process, called "pyrolitic" because the coating was baked onto the glass; or newer hybrid technology. Reflective metal coatings reject both infrared heat and visible lightwaves. They reduce glare for occupants and solar heat gain through

Glass *continued*

the building envelope. High performance coatings, especially the initial generation, consist of multiple coatings which can be damaged by oils in the human hand, and must therefore be captured on an interior face of insulating glass units, against the airspace. Low-emissivity coatings ("low-e") reduce the heat gain by rejecting infrared heat waves, but permit visible spectrum lightwaves to penetrate the series of metallic coatings on the glass, and therefore have a high visible light transmission and appear very clear, while offering improved thermal performance over uncoated insulating glass. These are typically high performance coatings requiring the use of insulating glass units, and generally have low outward reflectivity (often slightly less reflective in appearance than uncoated glass, even).

Tinted – body colored glass is available in a number of colors, primarily in greys, greens, and browns. Tinted glass absorbs heat and re-radiates out in both directions.

"Vision glass" is a term coined to describe the difference between glass which allows sufficient visible light transmission that the eye can see through, versus any of the above, alone or in combination, that might become impossible to see through. Generically, glass treated to become sufficiently opaque has been called "spandrel" glass, after the edge of slab spandrel beam condition where such opacity has historically been used to mask unsightly structural or other assemblies.

The heat absorbing aspect of tinted glass was an early treatment to help buffer the daily thermal cycle, used for instance in the Lever House, New York, Skidmore Owings and Merrill, 1952. Designed by Gordon Bunshaft at SOM, the building consisted of an open street level, a low podium building, with a rectangular slab tower floating above. The stainless steel-clad mullions spanned from slab to slab, forming a clean frame which was filled with vision and spandrel glass. The vision glass, Solex, was a body-tinted greenish colored glass. This tinted glass absorbed a percentage of the solar radiation but re-radiated the absorbed heat equally to both interior and exterior, effectively cutting heat gain by 50 percent. See Figure 4.30 for an exterior view of the restored facade. The original design included provisions for a roof-mounted gondola to clean the windows from the exterior, since the windows of this office tower were fully sealed, fixed glazing. Figure 4.31 shows the original section drawing which indicates the reach of the roof-mounted window washing rig.

Reflective coatings which subsequently became available reflect both visible and solar radiation, and were used on projects such as the John Hancock Tower, Boston, by I.M. Pei and Partners, 1976 (Figure 4.32), whose mirror-like facade was supposed to reflect the sky and clouds, thus rendering it invisible. The elegantly slim footprint and the coated glass both caused problems requiring post-completion remediation, the subject of much controversy and confusion, in part due to silencing orders from the subsequent litigation. In this case, the reflective glass was used in an insulating glass unit, combining reduction in heat gain with improvement in insulation value of the glazing.

Tempered glass, heat treated to improve the capacity of the glass to resist high local forces from through-bolt fasteners, or corner restraints, requires additional attention. Fabrication of any holes or shaping of the flat glass must be performed prior to the tempering process, where the glass is reheated and re-cooled for greater tensile resistance. However, the residual stresses may produce distortions and in some cases where low-angled lighting or polarized lenses are used to view and/or photograph the material, stress patterns may

◀ Figure 4.30

Skidmore, Owings & Merrill, Lever House – exterior view

▼ Figure 4.31

Skidmore, Owings & Merrill, Lever House – sections through the office tower show the range of the exterior window washing gondola, required to maintain the exterior since the windows were fixed

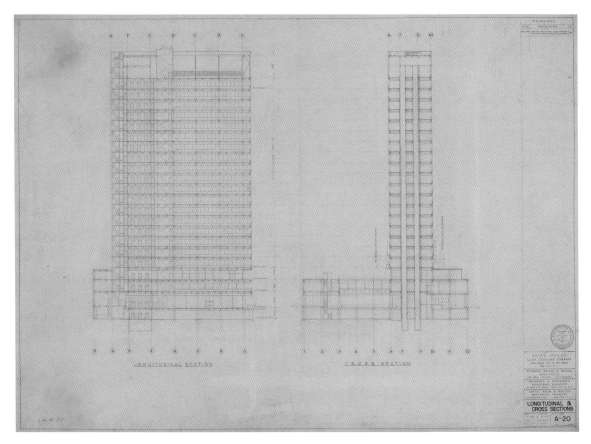

become visible and obtrusive. An example is Rafael Viñoly's Carl Icahn Laboratory Building at Princeton University, 2002 (Figure 4.33). This project used an outer layer of sun shades on motorized sensor control to shade the simple, single-seal glazing. The glass is mounted onto a series of cables, tensioned tightly against hidden structure above and below the glazed area (see Figure 4.34). These guitar strings, taut enough to resist the lateral pressures and suctions, support the glass with corner fittings (Figure 4.35), and are cross tensioned with horizontal cables where the glass joints also occur, but require no additional heavy backup in the zone of the glazing, resulting in a light, economical, elegant solution.

Windows and openings

Framed windows set into walls can be considered as a unit for infill into a wall opening (called the "masonry opening," if set into concrete, brick, or stone). The joint around such a unit would need to accommodate structural movements, thermal expansion, and any installation or erection tolerance. The gap around the opening would typically be sealed with silicone sealant, supported on an open cell foam backer rod, and possibly flashing if set across the gap in a cavity system. For a single-seal system, the integrity of each component at the outside face where rainwater might fall is the only protection against leakage. As discussed in the first chapter, the porosity of the constituent materials may result in vapor migration through a single-seal system. A two-seal cavity wall construction would then require that both seals maintain continuity against any infill element like a window unit. To achieve proper water protection, each element should lap over the

▲ Figure 4.32

I.M. Pei and Partners, Hancock Tower, exterior view

▲ Figure 4.33

Rafael Viñoly, Carl Icahn Laboratory, Princeton University, exterior view

▲ Figure 4.34

Rafael Viñoly, Carl Icahn Laboratory, Princeton University, circulation zone between sun shades and glass

◄ Figure 4.35

Rafael Viñoly, Carl Icahn Laboratory, Princeton University, detail of glass fitting on tensioned cables

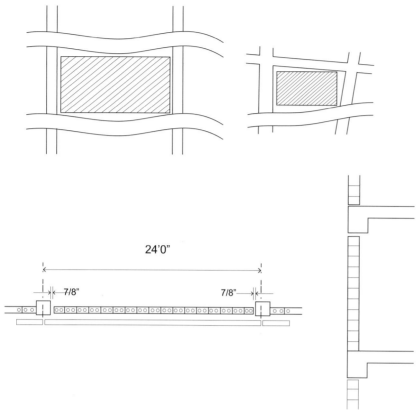

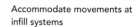

Accommodate movements at infill systems

next, shedding water outwards. This also means that at any opening, the infill unit should be detailed with overlapping flashing, end-dams, and closures as described in Chapter 2.

In the case of an exposed building frame with infill enclosure system, for example in industrial warehouses, the infill assembly responds to the forces of nature and transfers loads onto the building frame, which may itself experience global deflections from wind or earthquake, local loading movements, as well as some of the same temperature swings and external loads as the infill assembly. Loads imposed on the frame by the infill system and jointing issues for the components of each system are shown in Figure 4.36.

Historically, fixed windows were initially constructed from wood frames with removable stops holding the glazing material in place. One early method used in the West was glazier's putty, a soft, malleable, claylike material that held the glass in place. Setting spacers, called "glazier's points", were used in conjunction with the putty at the bottom of each lite of glass, to support the weight of the glazing, and were commonly made of lead so as to be soft enough not to break the brittle glass, but hard enough to transfer the dead load of the glass onto the frame. When this technology was translated into early metal windows, initially steel at the early part of the twentieth century, then aluminum, alternate materials were eventually adapted to provide the same cushioning, primarily gaskets, sealants,

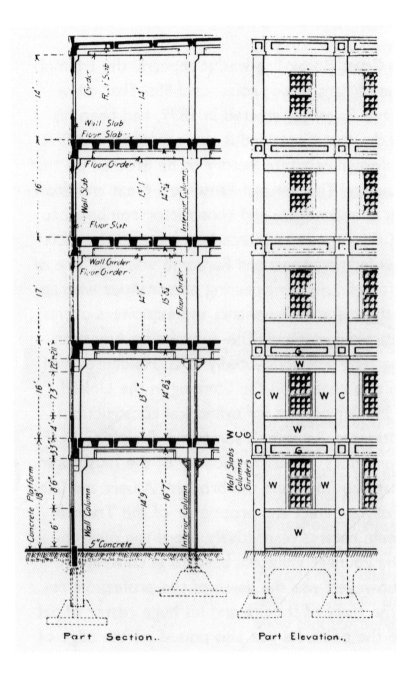

▶ Figure 4.37

Ernest Ransome and Alexis Saurbrey, *Reinforced Concrete Buildings*, 1912, rationalized formwork and construction into a kit of parts for reinforced concrete frame warehouses and industrial buildings

and elastomeric setting blocks. Like the glazier's points for spacing the glass within the frames, these materials must incorporate some flexibility to ensure that structural deflections and expansion and contraction from thermal swings would not constrict the glass or break it. Factories built to house increasingly large industrialized production lines often used cold-rolled steel windows in reinforced concrete structural frames, beginning in the early 1900s (Figure 4.37). The infill

◄ Figure 4.38

August Perret, Notre Dame du Raincy, exposed reinforced concrete in the design of a church, with stained glass captured in concrete

▼ Figure 4.39

August Perret, Notre Dame du Raincy, interior view showing nave interior

in these buildings incorporated extensive use of skylights and glass expanses to provide daylight in the early days of electric power. Limited by glass size and mullion design, and usually welded together from cold-rolled steel shapes, these industrial examples may not seem technically superior today, but inspired architects of the early modern era.

In part through the efforts of Auguste Perret, industrial materials such as reinforced concrete became part of the vocabulary of architecture, rather than merely vernacular industrial construction. Exposing the material to view, and appreciation, elevated the previously humble material, thanks in part to quality of workmanship of the wood form builders. At the church of Notre Dame de Raincy, outside Paris (Figure 4.38), Perret introduced the vocabulary of exposed reinforced concrete to formal ecclesiastic architecture. The interior view (Figure 4.39)

▲ Figure 4.40

August Perret, Church of St. Joseph, Le Havre, interior view of bell tower with stained glass concrete infill, much like Notre Dame du Raincy

▶ Figure 4.41

Auguste Perret, 25 bis Rue Franklin, Paris, France, exterior view

illustrates that the infill to the reinforced concrete structural frame was pieced together like a quilt from many small geometric shapes, repeated to create a modern stained glass infill. At the Church of St. Joseph, Le Havre, Perret later used a similar infill: the modules of glazed infill became mosaics of color, complementing the once prosaic concrete to create a soaring bell tower (Figure 4.40).

These churches and other seminal projects, such as the apartment block at 25 bis Rue Franklin (Figure 4.41), exploited the economies of the new technology of reinforced concrete construction in refined buildings. The efficient Hennebique concrete structure of the apartment building and the clever configuration of the street facade provided large window openings for daylight and air without loss of valuable floor plate since they eliminated the need for an interior courtyard. The Rue Franklin building frame provided a free, open plan that allowed placement of non-structural interior partitions wherever desired, introducing a spatial possibility that became important in the history of modern architecture. These projects also exposed the raw concrete structure to view, though they were decoratively treated at Rue Franklin (Figure 4.42), or refined by hand-finished surfaces such as bush hammered concrete infill areas and elegant tapered, fluted columns, at the Public Works headquarters, Travaux Publics.

◀ Figure 4.42

Auguste Perret, 25 bis Rue Franklin, Paris, France, detail of exposed reinforced concrete frame and infill

◀ Figure 4.43

Le Corbusier, Swiss Pavilion, exterior view

▼ Figure 4.44

Le Corbusier, Swiss Pavilion, exterior view

Perret shared his knowledge of concrete with Le Corbusier for his early Domino studies. In the Swiss Pavilion of 1930, Le Corbusier combined reinforced concrete construction (Figure 4.43) with the use of layered spaces and transparent facades to produce a conceptually new architecture. The first of two student housing projects he designed at the Cité Universitaire, the dormitory tower (Figure 4.44), used a streamlined exposed building frame with sleek infill of strip glazing and panels. In the reconstructed wall section (Figure 4.45), we can see that the primary reinforced concrete structure with infill facade was extrapolated in the later Unité d'Habitation projects such as at Marseille, illustrated in Figure 4.46, where glass and colored surfaces were combined in syncopated patterns as infill to the deep concrete framing structure.

A recent example of an expressed reinforced concrete frame with a house project is the Stone House in Tavole, Italy, completed in 1988 where Herzog & de Meuron used dry set stone slate as infill (see Figure 4.47). The spare lines of the primary concrete framing are exposed to the exterior and filled with rustic horizontal layers of stone with insulated cavity wall construction. As shown in the detail plan (Figure 4.48), we see that for design reasons, the primary corner columns were wrapped in the dry-set infill slate. An air gap separates the veneer

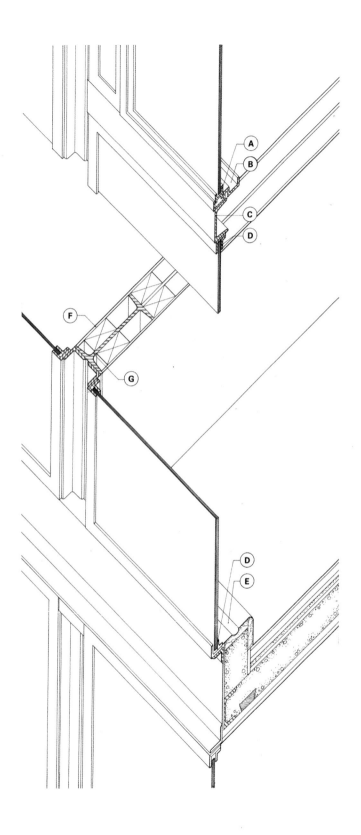

◀ Figure 4.45

Le Corbusier, Swiss Pavilion, axonometric showing wall construction. A – sliding steel window. The operable window consists of two horizontally sliding sashes made from rolled steel sections. The track in which they slide is formed from steel sheet braked (bent) into roughly a w shape to form grooves. B – track for sliding window. C – steel channel for support. In addition to forming the traditional rail 3 feet from the floor, this supports the considerable weight of the glass and the frame above. D – typical steel frame and stop, made from two rolled-steel angles (l-shaped pieces): a larger one on the exterior, forming the frame proper, and a smaller one on the inside, forming a removable stop for reglazing. E – sill with condensation gutter. F – typical partition. G – steel column, concealed within partition

◀ Figure 4.46

Le Corbusier, Unité d'Habitation, Marseille

▼ Figure 4.47

Herzog & de Meuron, Stone House, Tavole, Italy, exterior view showing exposed reinforced concrete frame and dry-set slate infill

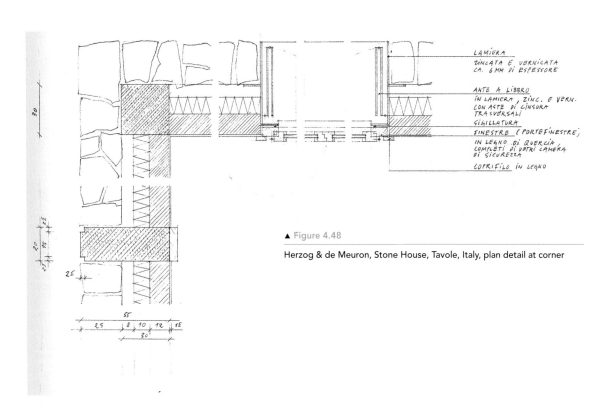

▲ Figure 4.48

Herzog & de Meuron, Stone House, Tavole, Italy, plan detail at corner

CHAPTER 4 Frame and infill construction

stone from the insulated inner backup construction. At the window openings, metal surrounds delineate a thin, clean edge for the inset windows and operable shutters. The dry-set slate of the walls blends into retaining walls of the garden and the open concrete frames outdoor space belonging to both the house and the landscape beyond.

Windows and operation

For ventilation, traditional operable windows give individual occupants local control of their comfort. The majority of windows typically used as infill in wall openings can be broken down into two basic categories: those with swing operation and those with sliding framed sub-units. Swing types include casement, awning, hopper, and jalousie windows. Sliding windows include single- or double-hung sashes as well as side-to-side operated sliders. The development of framing evolved from wood framing, which can be cut and shaped easily, but being porous requires frequent upkeep with painting and finish maintenance and even with regular care is vulnerable to rot and degradation. Glass was held in place using removable wood glazing stops to restrain the glass against the primary framing on all four sides. To some extent, adding layers of framed glazing is effective in improving thermal performance at the holes in the insulated building enclosure (the windows), though the introduction and widespread use of insulating glass units has reduced the need for multiple storm windows in more temperate climates.

Double and triple glazing has historically been used to improve the thermal performance of vision glazing. Initially, this was achieved by adding framed glass units within a single window opening. A famous example from Scandinavia, where extreme temperatures naturally made multiple glazing more common, historically, is evidenced in the design for a tableware piece by Kaj Franck, an award-winning Finnish designer. His Kilta creamer, designed in 1948 for the Teema Collection by Arabia, was sized to be narrow enough to fit neatly between the outer and inner storm windows common in residences at the time. The double-glazed outer window would allow the temperature gradient across the window assembly to keep the gap between the outer window frame and the inner storm window above freezing, but cool enough to preserve the cream from spoiling, even without refrigerators.[1]

Since the development of insulating glass units, double glazing has become common and triple glazing is increasingly available, even in the U.S.

Early steel windows, initially only hot-rolled simple shapes assembled into fairly crude capture channels, were strong and could provide relatively long-spanning members with thin profiles. Variations on these early shapes, both hot- and cold-rolled, are still available today. However, a large percentage of the commercial window market has turned to aluminum framing members. While steel window framing might originally have been putty-glazed, like some wood windows, gaskets have typically supplanted the putty, having a somewhat longer lifespan. Newer steel systems integrating hybrid aluminum glazing stops or beads have been used as well, though proper isolation of aluminum from

for a four-sided gasket frame. Naturally, this is only cost effective when large numbers of repetitive lites are to be glazed with the same gasket dimensions.

"Wet glazing", the term describing application of sealant in liquid form which then cures to a flexible semi-solid state, has the advantage of being reasonably reliable provided that the adjacent surfaces have been properly cleaned, primed, and prepared (different manufacturers and products have specific prime/prep requirements, sometimes as simple as clean water rinse and/or wipe with isopropyl alchohol, "IPA"). Gaps in sealant or other damage is readily repaired, as select areas of the material can be cut out and it typically bonds to itself reasonably well. It does, however, require a fairly skillful glazier to apply and must be pressed into the joint with a concave tool to ensure that the two sides being sealed achieve a good bond. Since it is a process installed on-site, wet glazing's effectiveness is dependent on workmanship and site conditions to achieve a good assembly. Problems that may occur, besides preparation of the substrates, include adhesion to adjacent surfaces, integrity of the sealant material itself, and proper joint sizing (see the joint sizing box). If sealant joints are not properly proportioned, roughly twice the depth of the sealant bead, the seal may fail internally to the sealant. If a joint is too deep, the sealant may not stretch sufficiently, and if too thin, it may not maintain integrity. If the joint is too narrow, even properly prepared sealant may also fail by being stretched past its movement capability (cohesive failure). In addition, all sealant joints should be prepared so that the two sides are the only surfaces receiving silicone sealant. When three-sided adhesion occurs, the sealant may also fail cohesively by tearing internally since the effective joint could be significantly smaller than the face dimension of the sealant.

Proto-unitization

The idea of infill construction, fitting between the framing or other exposed components of the building structure, evolved from industrial examples into the architectural realm. Warehouse and factory projects, like the Van Nelle Tobacco Factory, designed by Brinkman and Van der Vlugt (Figure 4.53), were able to make use of many repetitive elements for their exterior skin construction, as well as in the production lines the buildings would shelter. Because the size of these projects was sufficiently large to warrant some forethought to the production of exterior cladding systems, efficient manufacture and erection of the exterior enclosure for these industrial examples was required.

The modularization of Jean Prouvé's Fédération Nationale du Bâtiments, or Fédebat, Paris, 1949 (Figure 4.54), was made even more efficient by the production capabilities of his aluminum works, where entire panels were prefabricated complete with vent holes at the top, internal sash workings for operable windows, and opaque spandrel sills which hid the sash mechanisms, ready to be brought to the site and lifted into place. In the construction photograph, we see that the exposed edges of the building's concrete slabs and the recessed columns were cleanly completed, ready to receive the exterior units. In construction, a hand-operated rig was used to set the units onto a starter sill ready to

▲ Figure 4.53

Van Nelle Tobacco Factory, Brinkman & Van der Vlugt, 1925–31, outside Rotterdam. By cantilevering slabs out from the columns, the exterior cladding of this factory building was able to achieve a flush enclosure, hanging down to cover the slab with facade units that incorporated large expanses of glass to maximize daylighting potential

◄ Figure 4.54a

Fédération National du Bâtiments, Jean Prouvé, view of street facade during construction shows unitized construction of infill panels

▶ Figure 4.54b

Fédération National du Bâtiments, Jean Prouvé, view looking out towards street as infill unit is lifted into place

▶ Figure 4.54c

Fédération National du Bâtiments, Jean Prouvé, view A5213 shows unit being raised into place, and stockpile of units ready to be installed at lower right

▲ Figure 5.45d

Fédération National du Bâtiments, Jean Prouvé, view of garden facade

receive the modular panels, clearly a system that would minimize on-site construction time and maximize the precision of component assembly by allowing it to be put together in a very Taylorized, efficient manner, at the factory, rather than having to be executed in spite of any unfavorable field conditions at the construction site. This method of industrializing the construction process looks ahead toward the unitized curtain wall assemblies that would become more commonplace a half-century later.

Louis Kahn's Richards Medical Laboratories, University of Pennsylvania, 1957–1965 (Figure 4.55), also used an infill strategy for the design of its exterior

▲ Figure 4.55

Louis Kahn, Richards Medical Laboratory, Philadelphia, PA, 1957–65. View from corner shows brick cavity wall infill of exposed reinforced concrete building frame, and brick HVAC air distribution shaft towers, visibly expressed at the exterior

▲ Figure 4.56

Jørn Utzon, Bagsværd Church, Denmark, exterior

envelope. In this case, however, the exposed reinforced concrete frame defined square lab modules selectively filled with brick cavity wall spandrels and horizontal bands of glazing above. The mechanical shafts, expressed as brick towers on the exterior, form imposing elements off the sides of the primary structure, which cantilevers out in two directions at every corner.

Our last example, the small church at Bagsværd, just outside of Copenhagen, Denmark, by Jørn Utzon, 1974–1976, is likewise a concrete frame with very modest infill panels cladding the long side walls of the building, as seen in the exterior photograph (Figure 4.56). The stepped profile hints at the reinforced concrete shell structure spanning between the framed corridor structures. The interior view of the whitewashed concrete ceiling shows how it incorporates a light scoop transforming this very unassuming small town church into a warm, inviting congregation space bathed in clear light (Figure 4.57).

▶ Figure 4.57

Jørn Utzon, Bagsværd Church, Denmark, interior with altar and organ

CHAPTER 4 Frame and infill construction

Chapter 5

Cladding systems

Like infill systems, exterior cladding is a non-load-bearing exterior enclosure system that can be executed with a range of materials and assemblies, including the load-bearing systems covered in previous chapters. Non-load-bearing cladding skins resulted from the rise of reinforced concrete and steel as building structure. This created the opportunity to create larger, taller buildings whose structural stability and floor support system became an independent system from the skin enclosing the building interior. The skin or building enclosure system would therefore only need to support its own weight, without requirement for providing support to floor or floors above. Figure 5.1 shows a comparison diagram of infill versus cladding systems. The closure against the exterior environment still needed to shield the interior of the building from sun, rain/snow, and wind.

The development of larger, taller buildings was initially driven by commercial, industrial, and office uses. After the Great Fire in Chicago of 1871, the need for "fireproof" office and warehouse construction, and the rise of industrial production led to the need for buildings with larger footprints to accommodate larger machinery and equipment in increasingly dense environments in the most urban locations. These pressures combined with the evolution of building components we would now call a vertical transportation system (or elevators) and plumbing systems (bringing running water and toilets to upper floors) permitted the construction of taller and taller buildings. As these structures evolved, the exterior enclosure system protecting inhabitants from the outdoor environment grew thinner and lighter since they no longer needed to provide structural support. The infill systems studied in the previous chapter were used frequently in the early reinforced concrete frame industrial buildings and taller skyscrapers, but the structure was unprotected from the elements, in these cases. Though relatively massive from both a weight and thermal mass point of view, the unprotected structural slab edge or beams, and the reinforced concrete columns in these buildings were fully exposed to the weather. Though the concrete cover to the reinforcing steel is thick, the natural porosity of concrete would permit rain and moisture to penetrate to the tension-carrying steel, which would then rust and cause irreversible failure to the primary building structure. By covering the primary wind-force-resisting and load-bearing structure with a thermal and moisture protection, the exterior enclosure skin can prolong the useful life of the structure significantly. The separate cover, or cladding, can be assembled from any number of materials and systems.

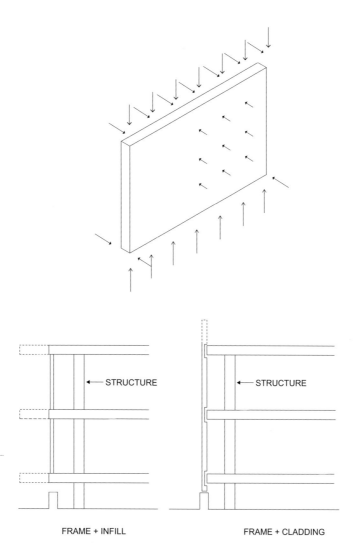

▶ Figure 5.1

Diagram illustrating the differences between infill construction and cladding or curtain wall assemblies, which are hung outside the edge of building structure

FRAME + INFILL   FRAME + CLADDING

Enclosure systems where the self-supporting enclosure system extends over to cover the surface of the superstructure have advantages over infill systems from a thermal standpoint, as well as for moisture, rain, and snow protection. Wrapping the entire building structure with a continuous barrier against wind and water in all forms, while also providing thermal insulation against temperature extremes in the outside environment as we will see, can be developed to incorporate increasingly sophisticated strategies to address and control the forces of nature.

Like the infill systems we reviewed in the previous chapter, cladding systems are only required to support self-weight. Loads imposed on the skin are transferred to the structural frame, in the case of structural loads (wind, weight), but the system is typically outboard of the structure and/or slightly engaging the structural frame. This way, the building structure is completely protected from the exterior elements and thermal loads. The same structural forces are transferred from the exterior to the skeleton as in the case of infill construction, but loading points might be quite different, depending on the size of the exterior cladding panels and their nature. While generally lighter than load-bearing systems, some exterior cladding can be larger scale with certain benefits (less erection time on-site, less labor on-site, longer spans).

Early examples of cladding systems include Otto Wagner's Post Office Savings Bank, Vienna, 1904, where panels of polished white marble are affixed to the backup structure at four corners using exposed rivets illustrating a non-structural cladding system (see Figures 5.2–5.6). A symbol of the modern, industrially sophisticated Austro-Hungarian Empire at the height of its power, the Post Office Savings Bank was, according to Ken Frampton, a "gargantuan metal box" due to the exterior facade expression and the use of aluminum throughout – glazed canopy, entrance doors, balustrade and parapet railings, as well as at the bolts on the thin, light "Sterzing" stone panels.[1] For the building skin, with its expression of the attachment points using clearly visible fixings bolting through the surface of the cladding panel, used decoratively almost like buttons in tufted upholstery, this bold adoption of the rivets required by early steel structures indicated an appropriation of the industrial aesthetic. Like masonry bearing wall precedents, rain and snow would wet the wall, due to the natural porosity and absorption of the stone. The exposed fasteners make clear, however, that the facing material is attached to the solid backup wall. This foreshadows the true cladding systems we will examine here, and highlights the problem of moisture infiltration into a wall assembly.

With most single-seal systems which rely solely on the primary waterproofing layer, the integrity of the system is only as good as integrity of the cladding material. In the case of the Post Office Savings Bank, the cladding panels were stone and, like modern stone or masonry construction, the outer cladding veneer surface is intrinsically somewhat porous. So long as the moisture infiltration through the more porous assemblies is taken into account in the design of the overall wall assembly, and care is taken to control and conduct condensation and moisture, there is no limit to the type of material that could be selected for a cladding system (cost and building codes permitting). Natural materials and man-made materials selected for use in a non-load-bearing cladding assembly need only be capable of spanning between support points, withstanding the environmental forces to which they may be exposed, and transferring these forces onto the primary building structure.

The size and scale of cladding elements may also be a function of the material selected, how it is made, or how it is transported. Besides the smaller modular brick sizes and dimensions tied to the human hand we studied in the bearing wall chapters, non-structural cladding panels can be as big or as small as can be fabricated, shipped, and erected onto the structure, so long as it is economical to do so. These lightweight screen cladding panels may not be possible to fabricate in very small sizes. Or, it may not be recommended because the material is costly to install in very small dimensions because of the size and cost of support framing. It may also be prohibitive to call for very large-scale cladding elements due to how a given material is created, fabricated,

▶ Figure 5.2

Otto Wagner, Post Office Savings Bank, Vienna, exterior drawing showing design intent

◀ Figure 5.3

Otto Wagner, Post Office Savings Bank, Vienna, photo of facade showing stone cladding, affixed to the building structure with exposed bolts

▼ Figure 5.4

Otto Wagner, Post Office Savings Bank, Vienna, photo showing cornice detail and stone cladding

▶ Figure 5.5

Otto Wagner, Post Office Savings Bank, Vienna, photo showing detail of fasteners, located approximately one quarter of the length in from each edge, following the running "bond" of the stone coursing at the base

▶ Figure 5.6

Otto Wagner, Post Office Savings Bank, Vienna, photo looking up at fasteners in stone

◀ Figure 5.7

Headquarters, Willis, Faber and Dumas, Ipswich, England, Foster & Associates, photo showing transparency in view looking outward through structural glass and fins

or transported. Glass, for example, is made from molten sand, silica, etc., and theoretically could be melted and formed into infinitely large or infinitely long pieces, but the weight of moving a panel, the capacity of cutting equipment or lifting equipment, or even the limits on roads, bridges, or tunnels in the transport of the cladding panels may limit the actual optimal size for a single piece used on a building.

Single seal exterior enclosure systems may perhaps be best understood in glass examples, since the transparency of several projects lets us see the mechanism of self-weight support against wind and gravity forces, and observe the separating line between interior and exterior clearly.

The Willis Faber and Dumas building by Sir Norman Foster, built in 1975, Ipswich, England, used a pioneering structural glass system where plates of glass were hung from the edge of the building structure, backed by perpendicular fins of glass used to brace the face glass against lateral wind loads (see Figure 5.7). By suspending the entire assembly from the roof slab edge compression forces were eliminated from the gravity support for the glass, thus producing the lightest possible configuration with clear face glazing and clear structural glass bracing. The uppermost panel of glass cladding, covering the third floor, was pre-drilled to accept bolts, which fixed each panel to the exterior of the structure at two points. A lower row of glass panels are suspended from the lite above, in turn using bolted connections to transfer the weight of each cladding panel through the upper lite back to the slab edge. At each intersection of glass, a patch fitting was used to transfer weight of glass, in this case using friction to develop a clamping action on the face of the bolted glass. The resulting transparency is evident in the night view (Figure 5.8), with only the thinnest possible profile of the stepped slab edge, barely visible inside the glass.

▲ Figure 5.8
Headquarters, Willis, Faber and Dumas, Ipswich, England, Foster & Associates, night view showing transparency of the glass skin and faceted curvature of the building footprint

Glass used for structural purposes obviously has become possible thanks to improved precision-drilling techniques and control of tolerances so it can be used as a structurally self-supporting component. The axonometric drawing of the Willis Faber Dumas building illustrates how the entire building section was configured to make use of the maximum glazed skin surface: the edge of the structure is minimized as the entire floor/ceiling assembly tapers and steps down to a thin concrete slab at the perimeter, maximizing light penetration to the interior. Careful design of the mechanical support and building systems distribution coordination is clearly required, alongside the use of the glass fins as lateral bracing against wind loads (see Figure 5.9).

Another example of glass used structurally is Sam Mockbee & Rural Studio's thesis project from 1999–2000 for Mason's Bend Community Center in Hale County, Alabama (Figure 5.10). In this project, the students in the design/build architectural studio

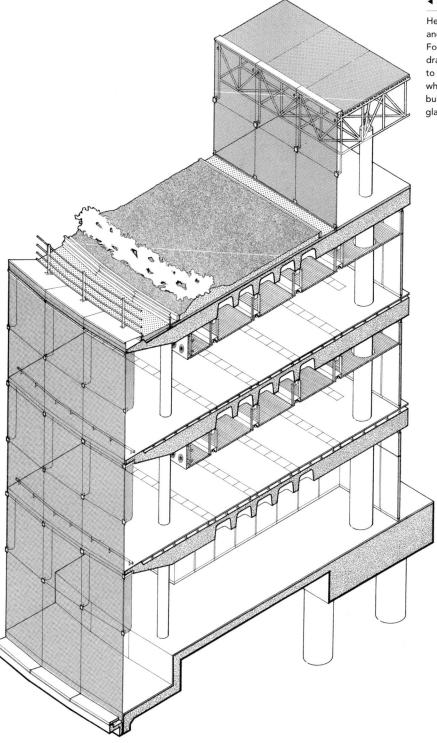

◄ Figure 5.9

Headquarters, Willis, Faber and Dumas, Ipswich, England, Foster & Associates, axonometric drawing showing stepped section to achieve slim edge expression while providing services to the building, and hung structural glass cladding

▲ Figure 5.10

Sam Mockbee & Rural Studio, thesis project, Mason's Bend Community Center, view from exterior showing shingled glass salvaged from automobiles

◀ Figure 5.11

Sam Mockbee & Rural Studio, thesis project, Mason's Bend Community Center, view from interior showing steel frame onto which the side car windshields were bolted, using the pre-drilled holes in the tempered glass

reused 1980s GMC sedan passenger side windows salvaged from a scrap yard, and bolted the glass panels onto a lightweight metal frame to create a structural glass shelter from inexpensive parts, shingled to shed water. Like the Willis Faber Dumas project, holes drilled in the glass were used to create attachment points for each cladding element. This example exploited the fact that side windows of automobiles are typically tempered safety glass, rather than front windshields which are usually required to have laminated safety glass for passenger safety during impact. The tempered windows break in a safe manner, and were drilled for the window operator crank prior to tempering. Thus the pre-drilled attachment points for the glass community building were available for fastening to the new steel frame (see Figure 5.11).

If the cladding material is a system comprised of smaller components tiled to cover the surface of the building envelope, a supporting backing surface or frame system becomes necessary to give more frequent attachment points for the smaller cladding elements. The entire assembly still ultimately transfers all of these loads back to the surrounding structure (see Figure 5.12). There is theoretically no limit to the size, small or large, so long as movements and expansions are taken into account and the weight of the cladding system can be handled by the primary structure within. Smaller cladding components result in a greater number of pieces, therefore potentially more erection time and labor on-site, and smaller spans from smaller cladding panels would result in shorter distances between support points (which could become more expensive), as well as potentially more joints/gaps/holes for introducing water infiltration (which could increase long-term leakage problem points). An illustration from stone cladding shows how smaller tiled cladding components are spanning from point attachments, which can be efficiently located one-quarter of the length of the panel, to take advantage of the offset from overturning of the end cantilevers (see Figure 5.13).

For some materials or assemblies, if they are large enough to span from one column to the next, the non-structural cladding can hang over the edge of the structure entirely and achieve certain economies of installation that look ahead towards curtain wall erection techniques. In both small and large cladding panels, joints between cladding panels become the next concern, after primary material selection. Regardless of panel size, it can be useful to think of each cladding element as sequins pinned at one point, ideally supported for gravity loads at two points, with edges as free or as fixed as necessary. Joints around these suspended panels are vulnerable to infiltration of water and air, so it can be wise to shape or profile the edges of the cladding panels to control moisture and water flow, at a detail level. Within the gap between panels, the horizontal joint surfaces perpendicular to the face of the building warrant a closer look. Some issues to consider incorporating into the joint detail include:

- Tilting the lower surface of the joint can allow water to flow back out and over the surface of the cladding panel
- Introducing a drip edge, reveal, or angled corner at the upper surface of the joint can prevent wind from driving rainwater into the joint and break the surface tension that could otherwise wick droplets into the building.

Like the overlapping clapboard shingles we examined in Chapter 3, the control of droplets and small quantities of moisture within the joint systems of exterior cladding systems can exploit the sloping lapped shingle principle to control water movement at joints and openings between panels. For large, fairly crude panels like concrete or perhaps stone, if fabrication costs permit, it is possible to shape or profile the edges of panels to incorporate these elements in the design of the cladding system joints. Vertical joints, where gravity would conduct drops of water down to the next horizontal deterrent below, can also theoretically be profiled in plan to deter capillary action. This offers one solution to minimize the suction of droplets from the exterior, where there is higher pressure

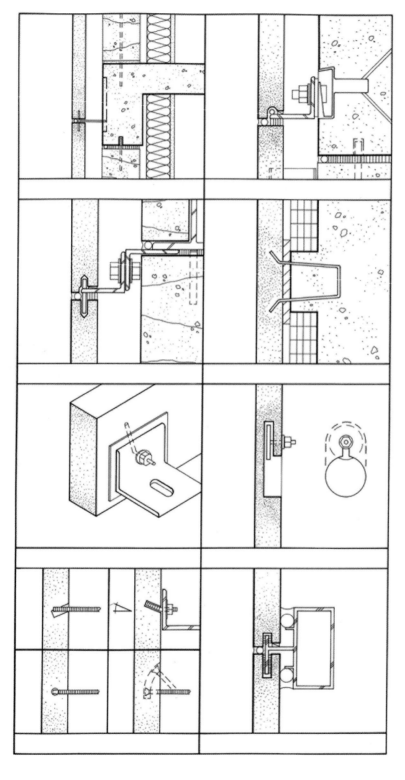

◀ Figure 5.12

Stone anchor details showing a number of proprietary options for achieving adjustability in all directions, to true out irregularities in structure

▶ Figure 5.13

Stone veneer cladding support positions, at quarter points of either horizontal or vertical stone cladding panels. Note that location of vertical control joints and horizontal joints should incorporate considerations of global building movements, thermal expansion, and live load deflections

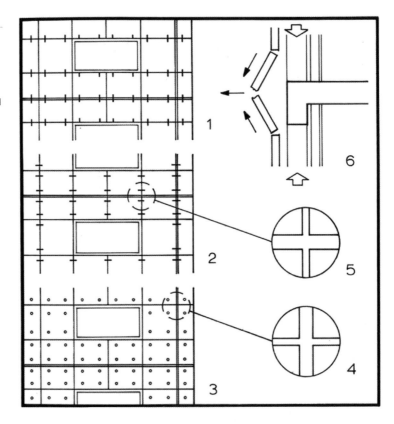

and lower pressure at the interior, or if the building is subject to internal rising suctions – also known as the chimney or stack effect. Horizontal joints would always ideally be profiled in section with a slight slope to permit flow down and out of any moisture or condensation. Horizontal surfaces risk the possibility of ponding, especially if construction or fabrication tolerances result in a dip or shallow puddle condition. With a slight slope it is possible to encourage water flow through a weep hole or gap in the outer seal, allowing controlled flow of water or condensation to the exterior. An even better detail could incorporate a stepped profile or gap, if possible, to create an internal edge preventing water infiltration to the interior, and to introduce capillary break, potentially. Profiling the upper surface of a horizontal joint with a sharp corner or sufficiently large reveal gap to overcome capillary action will again force moisture to drip away, as discussed in the masonry chapter.

Panels can be made from monolithic, poured materials that then harden into a stable stiff shape, like all concrete or cement-based materials, or any other products that can withstand extended exposure to sun, wind, and water. Precast concrete, reinforced with steel mesh, for example, and some glass fiber reinforced cement materials (GFRC) can be used on the exterior. Though somewhat porous, the vapor and moisture behavior can be incorporated in the design of an overall insulated skin assembly.

Precast concrete spandrel

It is also possible to build up cladding panels from a number of pieces, typically a frame support or truss structure with an over-clad panel, framing, glazing, or other relatively lightweight and stiff exterior finish assembly. Since individual panels can be manufactured off-site, the likelihood is that each of the panelized construction components can have greater precision in how they are fabricated at the smaller component scale, compared with the level of accuracy with which the building structure can be erected. Smaller tolerances can be expected with smaller building panels. Remembering the issues of joint sizing, this also implies that structural movement, construction tolerances, and thermal movements may introduce two scales for consideration. There are joints between the panel cladding components, with relatively small differential movements and loads, but the bigger the overall framed panel unit, the bigger the joints and construction tolerances between frames. Global building frame movements and tolerances, which are likely to be larger, must still be accommodated in the design and detailing between the independent panels on the exterior cladding.

A simple precast concrete spandrel cladding panel, as shown in Figure 5.14, can be fabricated off-site, trucked to the site, and lifted directly onto the building. As with tilt-up and other load-bearing construction in precast concrete, the large panels can easily be fabricated off-site. If they are designed to span column-to-column, live load deflections can be removed from consideration (see Figure 5.15), and infill windows can be installed within the concrete panels as well. Any configuration or panelization can be considered so long as it can be reinforced sufficiently to resist lateral forces on the final installation, as well as being shipped and lifted onto the building.

The thirteen-storey-high building at 290 Mulberry Street, a condominium building designed by SHoP Architects in New York, NY, used a series of prefabricated concrete panels faced with brick to achieve an unusual rippled effect (see Figure 5.16). By setting up a mold with stepped support for face bricks then casting concrete over the brick, the facade could be executed in large

Figure 5.14

Precast concrete outboard cladding spandrels installed on columns – Queens Hospital Center, 2000

▶ Figure 5.15

Full bay precast concrete panel installation illustrates how larger-scale components can be used to enclose the building rapidly, though joints between panels may be larger with increased unit size due to thermal and building movements

▶ Figure 5.16

SHoP Architects PC, 290 Mulberry Street, New York City. Bricks embedded in precast panels, stepped to create depth and shadow on the exterior surface

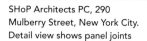

SHoP Architects PC, 290 Mulberry Street, New York City. Detail view shows panel joints

panels off site. Once mounted on the building, window units were set within the precast in a non-load-bearing assembly with considerable depth and shadow (see Figure 5.17). Though the appearance echoes the context of the neighborhood's largely load-bearing hand-set conventional brick buildings, the construction method is quite different and can be read in the joint lines at the face of the exterior skin and becomes apparent when the facades are viewed from either end, against the neighboring lots.

An example of a trussed panelized cladding system was used in the Highline 23 project by Neil M. Denari Architects, New York, 2011 (see Figures 5.18–5.19). In this project, prefabricated light framed panels spanned a full floor

▶ Figure 5.18

Neil M. Denari Architects, Highline 23, exterior view

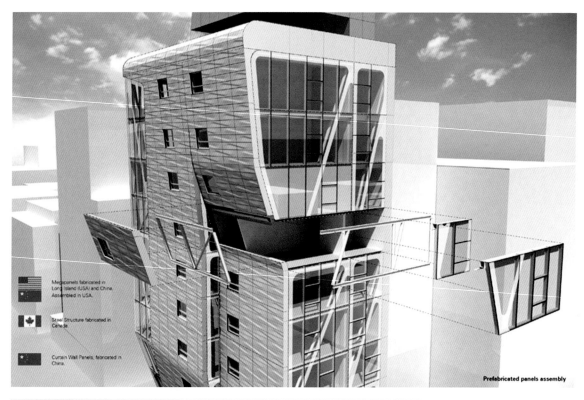

◀ Figure 5.19

Neil M. Denari Architects, Highline 23, architect's exploded view showing glazed and metal-clad trussed facade panels

◀ Figure 5.20

Neil M. Denari Architects, Highline 23, exterior view looking up at stainless steel truss panels with punched windows

height vertically, and essentially a full column bay in width. These large-scale panels were assembled off-site for both glazed areas and stainless steel panel units. Components pre-glazed in China were shipped to the site and lifted onto the building, as shown in Figure 5.20. Stainless steel rainscreen panels fabricated in Canada were brought together and assembled onto the large-scale trusses in Long Island City, Queens, New York City, trucked into Manhattan, and installed (see Figure 5.21). The finished rainscreen elevation, shown in Figure 5.22, incorporates fabrication and assembly joints for the smaller panels and moving joints between the larger truss units.

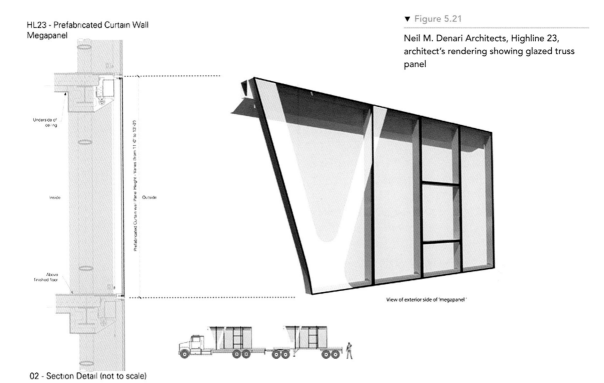

▼ Figure 5.21

Neil M. Denari Architects, Highline 23, architect's rendering showing glazed truss panel

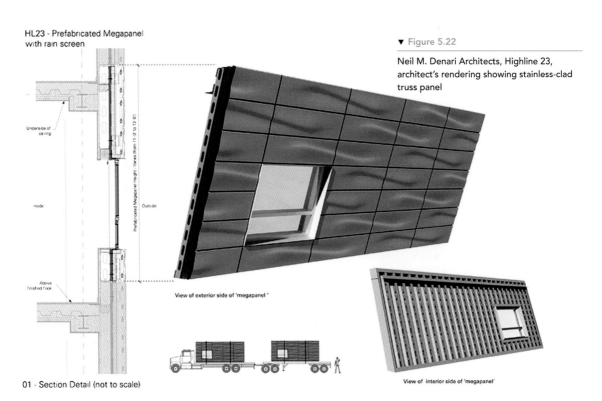

▼ Figure 5.22

Neil M. Denari Architects, Highline 23, architect's rendering showing stainless-clad truss panel

CHAPTER 5 Cladding systems

In the case of precast concrete panels or larger units, like the full bay units at Highline 23, joints between the large units require a flexible yet watertight closure. It is ideal to seal the components with two lines of defense, at these conditions. An even better improvement is to angle the materials forming the horizontal joints to further allow water runoff. Though concrete and most masonry and stone is somewhat porous, unlike the stainless steel panels used in the Highline Project, the design of gaps between them should still incorporate as many of the basic forces that drive water penetration as possible.

Rainscreen/double seal systems

The term "rainscreen" is used to describe a subset of cladding enclosure systems. A rainscreen system typically incorporates two lines of defense. The first line, or surface, is an outer water-shedding layer screening out direct exposure to sun, wind, and, of course, rain. The second, an inner backup layer, is typically watertight (see Figure 5.23). Having an outer screen shields the inner watertight

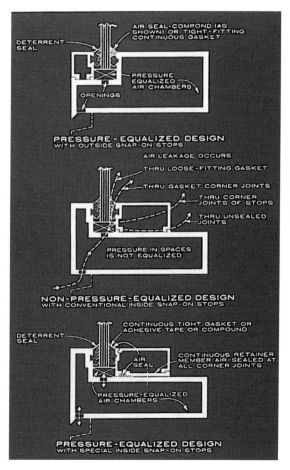

◀ Figure 5.23

Window sill section details illustrate how careful detailing of window profiles can incorporate rainscreen principles to achieve a pressure equalized chamber within extruded aluminum profiles in both out-swing operable windows and in-swing hopper windows. The stepped sill provides a lapped surface draining water droplets to the exterior, in both cases. For the out-swing window, the glazed window frame closes against the stepped sill with at least one continuous interior air seal gasket around the perimeter. Weeps or slots in the outer seal provide an entry point for air, which allows the chamber inside to achieve the same air pressure as the outside wind forces, thus achieving pressure equalization. For the in-swing hopper window, double lines of gasketing on the operable component could allow this detail to achieve two continuous seals. Slots or holes, which can also function as weep holes, allow pressure equalization of the interior chamber

enclosure from much if not most direct solar spectrum, both UV and visible light. The majority of infrared heat gain also hits the primary screen surface, further protecting the waterproofing interior surface. By relying on the interior waterproofing, the primary screen is relieved of some of the requirements and does not necessarily need to achieve a watertight enclosure itself. This dual-seal strategy can be used to relieve the assembly of requiring perfect workmanship in the exposed outer system. While it does put greater emphasis on the backup to maintain the watertight, fully sealed skin separating interior from exterior, that backup surface often does not need to be aesthetically perfect so long as the performance is achieved since the outer screen material obscures most, if not all, of the secondary surface from view.

Advantages over single-seal systems

Since a rainscreen system does not need to provide a fully watertight outer enclosure, it is possible for it to be assembled from a series of cladding panels with partially or fully unsealed joints. Such "open joint" systems offer the advantage of displaying crisp, clean edges of each panel with a darker shadow at the joints, rather than the smooth, uniform surface that would be produced by closing joints with sealant color matched to the panel color, or a grid appearance if a contrasting color joint sealant were to be installed. Even with an appearance of being fully sealed, it can be possible to employ rainscreen principles by introducing sufficient gaps at regular intervals – these gaps could be incorporated into the design of the condensation or weep system, and would still offer the workmanship issues described above, where a primary seal and primary surface help buffer the interior seals and relieve the interior watertight surface from having to achieve visual perfection as well as provide watertight performance.

Single-seal systems are simpler and may be more accessible, for example if the building they are used to enclose is lower in height. Resealing from the exterior can be very straightforward in a single-seal system, by cutting and removing the original joint filler material and replacing with new. Dual-seal systems are likely to be harder to maintain, especially if the hidden secondary waterproofing seal surface is compromised. Removal of the outer screen to access the inner surface would likely be required, unless access from the interior could be achieved without significant destructive work. While it can be very advantageous to incorporate sloped, shingled, or nested profiles within each layer, replacement and remediation in future years can be quite difficult. In addition to these technical difficulties, a dual-seal cladding system can be more expensive than some single-seal systems. In the case of a perforated outer rainscreen panel, for example, achieving an interior watertight enclosure surface becomes more difficult and costly since the inner surface is also exposed to sun, wind, rain, and the elements, at least in selected areas. Though it can be very effective from a heat gain standpoint, for instance, where the outer screen shields the interior enclosure system, such a building would have practically two independent enclosure systems and, potentially, additional cost for two such systems. In locations where temperature and climates are extreme, it may be well worth the initial cost since long-term heating/cooling loads would be significantly improved, so a life-cycle energy study should be considered in conjunction with the architectural design.

Cladding materials

Most wall construction types can be used as rainscreen or cladding systems, including masonry, wood, stucco, EIFS, precast concrete, metal, stone, and glass. How each material is fabricated and supported to become non-load-bearing, from a global building structural sense, can of course vary. Some materials, like metal and glass, are industrially produced and therefore only limited by how they are made. Glass, being a floated production, is often limited more by coating, fabrication, or lifting tolerances than, for instance, metal, which is lightweight and easily lifted, but is only commonly available in certain widths and can be liable to buckle without adequate stiffeners backing it. Natural materials, like stone, are again subject to strength calculations if used in thinner veneer slab shapes, both for the panel themselves acting as a beam against support points and locally at fastener anchoring conditions. Precast panels are not typically constrained by either production limits, fabrication constraints, or material strength (if reinforced), but lifting weights and shipping sizes can be the determining factors. Smaller components can be assembled into larger-scale rainscreen or cladding uses, as well, provided that the method of support and anchoring do not compromise the continuity of the inner watertight line of defense.

To begin a detail study of how rainscreen principles can apply to exterior cladding systems we could look to previous masonry cavity walls and see how they can be used as a non-load-bearing exterior cladding system. In some interpretations, a cavity wall could be considered a sort of rainscreen. In our typical cavity wall where the interior structural wythe rests on the structural slab while the veneer masonry is supported outboard of the line of the structure, we could consider it to be incorporating a primitive version of a pressure-equalized rainscreen cavity to resist water infiltration; flashing and weeps can be used to control water movement and conveniently could also be used to delineate floor-to-floor movement joint locations; cladding materials and their joint filler materials can be designed as either fully or partially sealed so long as a backup secondary waterproofing is used to protect the interior.

Location of insulation and vapor barrier

Insulation within such a masonry cavity wall would be situated so as to be faced with an impermeable vapor-proof barrier on the warm side. In our cool, heating-dominated, northern hemisphere, the warm side of the insulation would typically be the interior face of the insulation. As shown in the section drawing and detail images (Figure 5.23), the dashed line of the vapor barrier maintains continuity around the entire building, tight to the interior face of the insulation even as it wraps around corners and intersections. The condensation collected against this vapor barrier is conducted to the exterior through the flashing system and weep holes, as we saw in a previous chapter. In a warm, tropical climate where outside temperatures are consistently warmer than air conditioned interior temperatures, throughout the majority of the year, this insulation and vapor barrier condition would be reversed. The condensation and moisture path would be reconfigured so that water droplets were still conducted safely to the exterior, minimizing opportunities for mold or other growths. Attachment points through the vapor barrier must be fully sealed to avoid compromises to the waterproofing of the entire assembly, though local points of penetration may be acceptable. Thermal breaks at anchor points for individual panels may not be possible in all conditions, though each project, climate, and situation may dictate whether this is acceptable. Stone anchors, for example, may need to be firmly bolted through a waterproofing vapor barrier surface. Though sealants and product systems may provide watertightness and offer warranties on protection, some thermal bridging in the diameter of each through bolt may be inevitable.

Detailing concerns

In the detailing of cladding and rainscreen systems, a number of issues need to be considered. The main building frame, or skeleton, is likely constructed of cruder, large-scale components, whether steel or rough structural concrete. The structural system typically is constructed in such a way that the final product is within an inch or two in every direction of the target building geometry with respect to floor elevation, position in-and-out of the face of the building, as well as side to side. On top of this, structural movements must be incorporated into joint and panel detailing. As with infill systems, movements, tolerances, and warpage should be designed into the skin system based on worst case assumptions and calculations for the primary building frame, in order to achieve the best looking and best performing final product.

To avoid past mistakes where movement was not sufficiently accommodated and uncontrolled random cracking occurred, modern architectural construction detailing pre-calculates these movements and attempts to design joints to permit these movements to take place without irreparable damage to both design and building integrity. Knowing that the precision of the architectural cladding or rainscreen system will be constructed to more stringent tolerances, it becomes important to incorporate mechanisms for adjustment in all three dimensions to ensure that the final building is constructed accurately and falls within property boundaries, in accordance with an architectural design intent. For some of the more technically sophisticated cladding products requiring extensive off-site prefabrication, it also becomes critical to predetermine the final building component sizes and shapes. Materials requiring off-site manufacture and assembly such as stone, metal panels, or other pre-made cladding elements typically involve a longer process of formal correspondence between the construction team and the architect, as subcontracts are selected and awarded, shop drawings are prepared, and fabrication begins, much less at the final installation on the building frame.

Sequence of installation

To consider an example using a variation of the brick cavity wall we could imagine a system with a spandrel cladding of brick veneer masonry. There is flashing above each supporting shelf angle. The shelf angles are anchored off the edge of the concrete floor slab with an assembly that has adjustment capability. The anchor is adjustable in the up/down direction because the insert can slide along the embedded groove. The veneer brick will be set with the hinged front wire in the bed joint at whatever position left/right and in/out is necessary, providing two other axes of adjustment to isolate the desired veneer from inaccuracies in the concrete construction. Note that there is silicone sealant and a soft foam backer rod under the shelf anchor, which turns the brick supported on the next shelf angle below into a panel, isolated from deflection of the structural concrete above.

The sequence of installation starts with the construction of the foundation and building structure. Once the skeleton is erected, the enclosure of the building becomes the next phase. The attachment system or anchors are attached to the skeleton, if not poured at the time the concrete is placed for the floor slabs. Critical coordination then must be sequenced to create the desired final system with proper lapped drainage: installation of waterproofing over backup construction, attachment of insulation, setting of cladding panels, and sealing of soft joints. The following individual steps are required in setting exterior cladding:

Step One: Adjustable anchors need to be set to receive the veneer masonry. The adjustment in x, y, and z axes relative to the theoretical face of structure allow the exterior enclosure anchorage to adjust to the desired position for the skin system and accommodate inaccuracies inevitable in the rougher construction of the structural building frame. This rather simple set of connections is a precursor to the increasingly sophisticated anchor systems used in curtain wall systems as we will see in the next chapter.

Step Two: Installation of flashing. Timing of flashing installation must be coordinated with and set on top of the embedment to create the anchor shelf for the veneer cladding in this example. This becomes more complicated since multiple trades and subcontractors may be involved, requiring sequenced work schedules.

Step Three: Installation of waterproofing. Care should be taken to ensure that it is lapped over the flashing to form a continuous waterfall of overlapping umbrellas protecting the backup and interior from moisture migration. For the entire assembly to function optimally, the system depends on the interior waterproofing surface to form a continuous vapor barrier.

Step Four: Installing insulation and vapor barrier. In the case of rigid insulation, thick foam sheets can be affixed to backup construction using an adhesive that can also be used as a vapor barrier, if applied properly in accordance with proprietary system warranty, in some cases. The vapor barrier should always be installed at the warm side of the insulation, so becomes dependent on an evaluation of the local climate. Locating insulation and/or a vapor barrier at the interior is sometimes the only option, though a careful review of the details can minimize conditions and configurations that might compromise the continuity of a vapor barrier.

Step Five: Installing veneer. The skin material exposed to view in the final installation, or veneer, such as one wythe of brick. All cladding panels and skin materials that are resistant to exterior loads and exposures can be used as a substitute, though their support requirements may be significantly different.

Step Six: Installing joint filler. Soft, compressible, and expandable fillers can be used to create closure at the gaps around veneer/cladding panels. Silicone sealants, if appropriately sized, can compress, expand, and remain adhered to the sides of the cladding while still providing a barrier against water and wind. Openings to allow drainage of condensation from the vapor barrier out through the cavity must be maintained in the form of weep holes or through drainage tubes.

Lightweight backup

There are some occasions where a lighter backup to veneer masonry is unavoidable, whether due to cost or weight concerns. Lightweight metal studs with exterior sheathing and/or some variation of insulated cavity construction is technically achievable in both infill as well as in a cladding configuration. In any case, the backup system should be rigid and provide sufficient support as well as stability to prevent overturning. Conduction and condensation paths should also be studied carefully to ensure that the interior components are protected from cold bridging wherever possible. Care should be taken to consider thermal performance of the system to ensure that the metal is insulated from exposure to the cold exterior by providing insulation outboard of the sheathing, ideally, with warm-side vapor barrier and condensation control plus weep paths.

Spandrel systems

Another variation on non-load-bearing cladding systems hung outboard of the primary building structure is the spandrel with infill system. In these examples larger-scale single components can be employed, like precast concrete panels or other large spandrel panels built up from smaller elements onto truss-like exterior enclosure panels of comparable scale/size (brick or stone on truss backup). For these cases, a significant fraction of the fabrication can be executed off-site, to minimize time on the construction schedule on the job site. This becomes cost effective in areas where job-site labor is expensive or a time-critical project is being undertaken, to reduce an overall construction schedule. Some of the variables to consider include balancing the cost of setting individual components onto a backup truss off-site, plus shipping prices (could be oversize load) in order to limit installation time and/or costs on-site. From a design standpoint, it is worth noting that larger panels carry larger joint requirements because of construction, fabrication, and erection tolerances which must all be added to determine the final joint dimension.

Examples of non-structurally load-bearing cladding systems that bridge between infill and cladding include early industrial factories and warehouses. In most of these examples, fairly extensive use of glass and daylighting was common. It is likely that the need to minimize costs in the early days of electricity, coupled with the practicality of maximizing maintenance-free lighting without any bulbs or lamps to replace, or wiring to create sparks around valuable machinery and production, helped drive the form of these buildings. Reminiscent also of the fireproof construction principles learned after the Great Fire of Chicago, the use of noncombustible concrete, glass, and masonry became prevalent in these pragmatically utilitarian buildings to protect the owner's investment in materials and machinery, if not the laborers as well.

In New York City the Starrett-Lehigh Building, designed by Cory, Cory, and Matsui in 1931, on West 26th Street, was also constructed by the Starrett Investment Corporation and the Lehigh Valley Railroad for commercial and industrial purposes (see Figures 5.24 and 5.25). Handling large volumes of materials and goods in the more constrained city context meant that, unlike the Boots Factory in England, perimeter vehicle access on all sides of the block-sized structure was not possible. However, an extension of the railroad extended into the lower levels of the building (see Figures 5.26 and 5.27), so materials could be shipped in easily and goods could be sent out of the city readily. In addition, truck entry dropped below the train track levels and allowed vehicles to enter an oversized elevator car, lifting them to upper floors where stepped loading docks were set into the reinforced concrete floors for loading/unloading locally at each tenant's space. This became increasingly popular as the bridges and tunnels around the city in the post-World War II era were completed and truck transport became the norm. The Starrett-Lehigh building was clad in a modern infill skin with exposed reinforced concrete slab edges. Horizontal strip windows rest on brick knee walls, set on a stepped exposed edge of the floor slab (see Figure 5.28).

◀ Figure 5.24

Starrett-Lehigh Building, New York City, exterior view from 1932

▲ Figure 5.25

Starrett-Lehigh Building, New York City, construction view from 1931 shows opaque spandrel panels being installed as soon as building structure is complete, with strip infill glazing following behind in the construction sequence

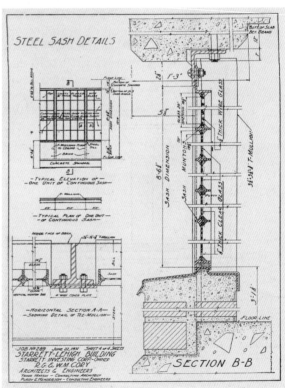

▲ Figure 5.26

Wall section detail showing operable strip windows. These infill windows incorporated a slip joint at the head as the original Architectural Forum article from 1931 notes, "Detail of specially designed sash, in which provision was made to offset movement of the slab"

▶ Figure 5.27

The Starrett-Lehigh Building was designed to incorporate transfer of goods between the Lehigh rail spur, warehouse space, and vehicular traffic. Loading and unloading was designed to occur inside the building, which occupies a full city block. Trucks enter mid-block off the street, travel below the railroad tracks, and enter heavy duty elevators which carry them up to the floors, which were stepped making loading docks at the floor to accommodate up to nine trucks on each level. They would then re-enter one of the truck elevators, without having to turn around, then descend to the street

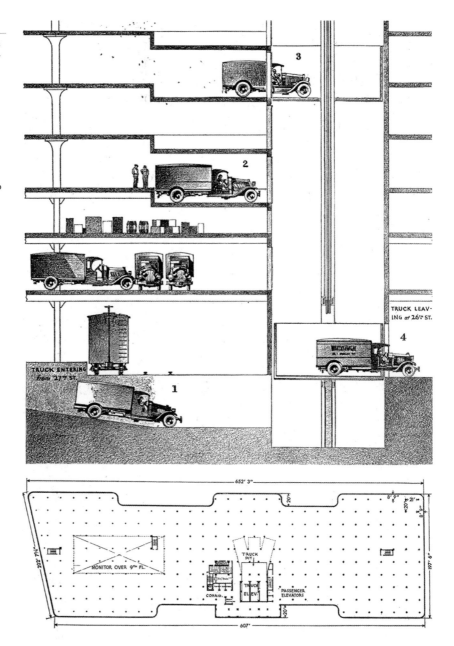

From these early industrial prototypes evolved such post-World War II streamlined successors as Frank Lloyd Wright's Johnson Wax office building and research tower, of 1950, built in Racine, Wisconsin. General views of the Johnson Wax complex show the tower rising above the low building and parking structure (see Figure 5.29). In the case of the administration building, masonry spandrels were used at the perimeter of the low structure with infill strip glazing under a large open forest of "mushroom" columns, an innovation by FLW where each circular mushroom touched the next for lateral stability (see Figure 5.30). No beams were needed, so this organic structural module was able to minimize

 Figure 5.28

Starrett-Lehigh Building, New York City, exterior view showing spandrels and infill glazing

▶ Figure 5.29

Johnson Wax Corporation Building, 1525 Howe Street, Racine, WI, Frank Lloyd Wright, 1950, exterior overall view showing spandrel and infill construction at research tower in the distance

▶ Figure 5.30

Johnson Wax Corporation Building, 1525 Howe Street, Racine, WI, Frank Lloyd Wright, 1950, interior view of administration building shows umbrella columns and perimeter clerestory

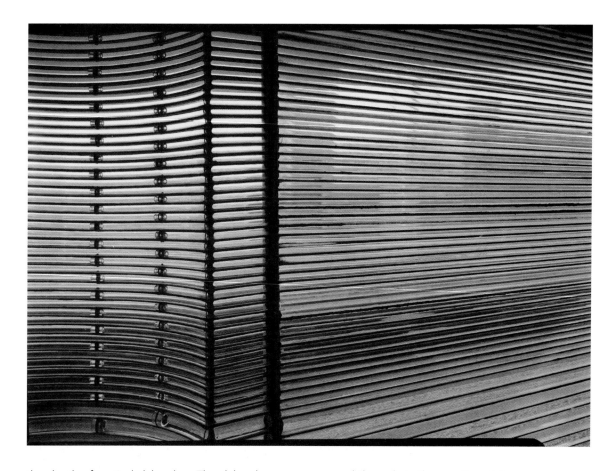

▲ Figure 5.31

Johnson Wax Corporation Building, 1525 Howe Street, Racine, WI, Frank Lloyd Wright, 1950, detail view of Pyrex tube glazing, used throughout

the depth of vertical slab edge. The slab edges are expressed through to the exterior, forming a contrasting horizontal stripe in white, like the two preceding commercial and industrial examples. Brick cavity walls were set on the edge of these slabs, further accentuating the horizontality of the streamlined aesthetic. A peculiarity of the Johnson Wax Laboratory and Administrative buildings was the infill skylights and infill glazing created from stacks of Pyrex tubing, stabilized by custom castings, and "glazed" or sandwiched between an early type of resilient gasket (see Figure 5.31). At the research tower, slabs at multiple elevations were cantilevered off the central core, in a variation of the mushroom pods used in the administration building, with spandrel brick masonry and white limestone highlight strips, continuing horizontal expression. The Pyrex tube glazing represents an interesting and rather clever use of sealed tubes to create an insulated barrier between interior and exterior. Since the insulating potential of captured air is maximized by creating a sealed cavity, the tubing with many ends and many individual joints proved vulnerable to leakage, and the castings offered no means of heat- or cold-bridge isolation. However, the masonry knee walls appear to have used double layers of face brick, offering some level of thermal isolation and filled with rebar as well as mineral wool insulation, in a woven grid of reinforcement (see Figure 5.32).

▶ Figure 5.32

Johnson Wax Corporation Building, 1525 Howe Street, Racine, WI, Frank Lloyd Wright, 1950, exterior of research tower base showing horizontality of limestone feature strips trimming the brick cavity wall spandrels

Spandrel cladding in particular and exterior cladding in general can also be executed with panels of other materials besides masonry, such as precast concrete. Larger-scale panels are quite reasonable to produce, provided shipping sizes and transportation limits are respected. An advantage of larger and longer panels, such as with precast, is that they can be wide enough to be supported off building columns, rather than only the slab edge, as we saw with the unit masonry examples above. By spanning column-to-column it becomes possible to mitigate the cruder fabrication tolerances of concrete by limiting the

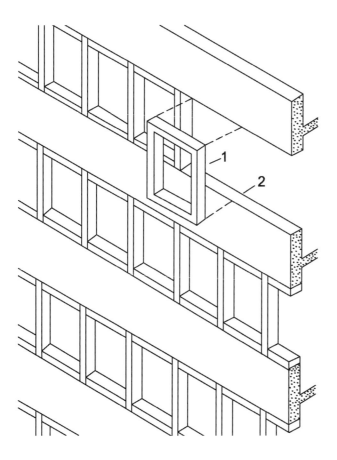

◀ Figure 5.33

Spandrel and infill diagram, AAMA
1 Infill
2 Spandrel

effect of structural deflections on the joint sizing between these components. Figure 5.33 shows a typical bay configuration for a panelized precast cladding system illustrating both types of precast concrete panel anchors – dead load as well as lateral bracing against wind. It is not uncommon to use the edge of the building slab for one or two additional lateral bracing anchor points, since the precast panel in this case spans column-to-column and can be thinner if there are additional wind anchor point(s) minimizing the required material strength and thickness.

Anchors can easily be incorporated within the casting process. Anchor components, whether rods or rolled steel profiles, can be positioned within the mold before the concrete is poured to provide attachment points for gravity load support, wind load anchors, and lifting brackets (see Figure 5.34). Lifting and installation loads for these large, relatively heavy panels may present the worst case torsion or point-loading scenarios that the cladding will undergo. Crane lifting capacities and shipping issues can also govern size limitations on panels – depending on efficiencies of construction erection, it is cost effective to minimize the number of individual pieces requiring an individual "pick" to minimize overall labor costs. However, weight limits on cranes (and associated costs during construction) may counter the need to minimize the number of panels,

▶ Figure 5.34

Precast concrete spandrel panel shipped on flatbed truck to project site with integral anchors cast into panels

and crating/transport/shipping limitations may also restrict panel sizes. These are all considerations that may affect design decisions.

Anchors for large panels can also be composed of two parts to accommodate adjustment, both likely steel shapes for strength. The brackets can be welded to the steel building frame, as shown in Figure 5.35, in preparation for precast installation, or could have components cast into the floor slabs. If loaded to the building columns, the only movements and tolerances besides fabrication of the precast panels themselves to consider would be shrinkage, deflections, or building-wide movements. This permits the smallest possible horizontal joint sizes. If the panels are attached to the building columns, the precast is effectively spanning like a beam, and sees no live loads from the primary building structure and no mid-span deflection from the floor slabs. Shrinkage of the

▲ Figure 5.35

Anchors on columns ready to take dead load of precast concrete spandrel cladding. Supplementary intermediate wind anchors are visible at the underside of each perimeter beam, mid-span between columns

building frame would be considered if the building frame were concrete as well, but in the case of a steel framed skeleton, the effects are negligible.

Wind anchors are not required to bear the weight of a cladding panel and can in some cases be simple embedded threaded rods, fastened to steel angles at the edge of the slab. The weight of the precast panel can also be transferred to the building structure on the slab edge, but so close to the columns that there would have been minimal slab deflection. In all permutations, the combination of anchors must adjust for fabrication, installation, and field conditions to allow for the worst case in all three axes to establish a finished position that is able to accommodate for the inaccuracies anticipated.

Panelized cladding can be prefabricated off-site from smaller cladding elements, such as stone, metal, or any other panelized weatherproof material, by mounting these components onto a large truss. The truss can be attached to a building much like the precast examples, with all of the positives: faster site installation of fewer pieces; controlled fabrication and assembly of smaller elements in a factory condition off-site rather than rough outdoor work on-site with attendant time constraints for project timeline; and less on-site time. Negatives include the crudeness factor of manufacturing and erecting larger panels, which likely implies larger joints; unwieldy pieces requiring larger lifting capabilities and probably larger cranes; less adaptability to unforeseen field conditions. It is also likely that the cost of the truss as well as the cladding materials could be too high for the positives to outweigh these negatives, steel being costlier than concrete, and the cladding materials likely being more costly than concrete. The strategy is worth remembering, however, in conditions where rapid site work is important, whether because of high labor costs, expensive real estate considerations, availability of skilled workforce, or some combination of all three.

There are also many proprietary systems that can accept lighter weight exterior finish panels, typically much smaller than the heavy precast examples. Attachments onto an intermediary rail system, horizontal or vertical, recall the systems we saw in some of the earlier masonry support systems. All incorporate some combination of pre-engineered components to provide a predictable surface that can adjust to accommodate inaccuracies in cruder systems, and three-way adjustability typical for most cladding systems. This is useful for attaching cladding materials to new building skeletons, but is also a good approach for re-cladding and repurposing existing buildings for new uses. Refined tolerances generally improve as the size of the component gets smaller but the methods of manufacture and the materials used can, of course, affect each element

Any number of materials and attachment systems could all be categorized as non-load-bearing cladding systems. The Sainsbury Centre for the Visual Arts, Sir Norman Foster, 1978, Norwich, England, used a large span truss as the structural frame for the building, to give a column-free space (see Figures 5.36–5.38). The clear glass and solid panel cladding system is supported on a grid at the exterior as well. Close-up photos of the exterior show the ribbed panels and duct penetrations incorporated into the grid in a rational system. Axonometric

▶ Figure 5.36

Sainsbury Centre for the Visual Arts, Sir Norman Foster & Associates, used a kit-of-parts cladding system

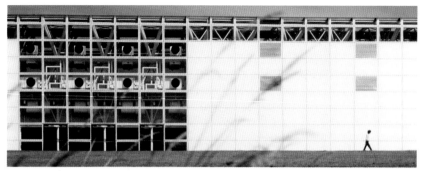

◄ Figure 5.37

Sainsbury Centre for the Visual Arts, Sir Norman Foster & Associates, elevation view of glazed panels at vision areas. Air intake and exhaust locations were integrated into the design and exposed to view

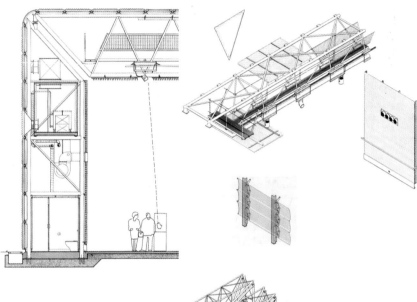

◄ Figure 5.38

Sainsbury Centre for the Visual Arts, Sir Norman Foster & Associates, axonometric views diagramming kit-of-parts cladding

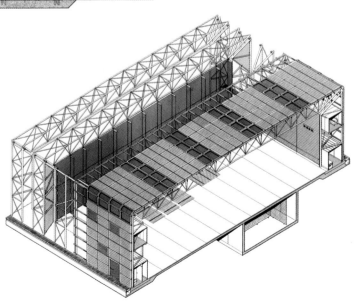

views of the building show how the cladding attached to the primary structure and the integration of the building systems with the envelope design.

At the Kursaal cultural complex in the Basque town of San Sebastian, Spain, designed by Rafael Moneo and completed in 1999, a series of prismatic volumes is enclosed in a double skin suspended off a hidden inner truss. Forming the enclosure for the convention, cinema, and concert hall, a double wall was built to bring light into the public lobbies by day, illuminating both the interior and exterior surfaces at night. Flat translucent glass was used at the inner layer, and curved laminated slabs of textured cast glass were held in another framing system, at the outside of a steel framed truss. The entire volume is lit by hidden fixtures at night, illuminating the interior and glowing from the exterior, reminiscent of how frames of moving film are back lit without being projected onto the big screen (see Figures 5.39–5.42).

▶ Figure 5.39

Kursaal, Rafael Moneo, San Sebastian, Spain, exterior view showing translucency of exterior glazing

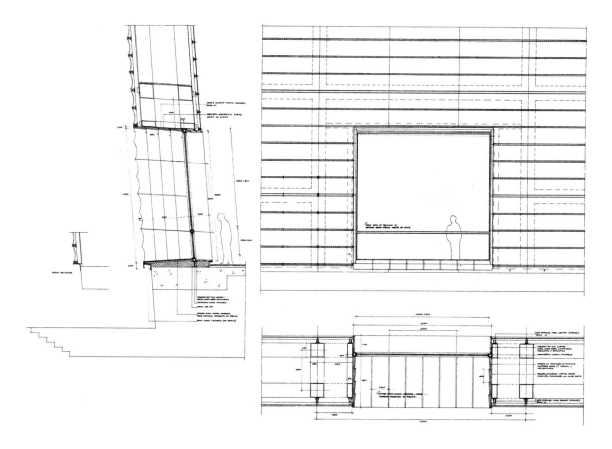

▲ Figure 5.40

Kursaal, Rafael Moneo, San Sebastian, Spain, plan, section, and elevation details showing double exterior wall assembly

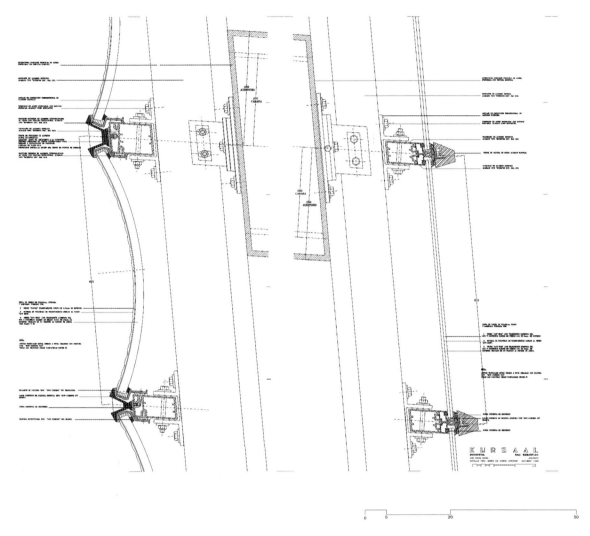

▲ Figure 5.41

Kursaal, Rafael Moneo, San Sebastian, Spain, section detail showing curved laminated textured glass at exterior, flush laminated glass at interior, supported on extruded aluminum profiles and steel frame, sandwiching lighting

◄ Figure 5.42

Kursaal, Rafael Moneo, San Sebastian, Spain, interior view showing muted daylight through glass cladding

Chapter 6

Curtain walls

Curtain wall systems are a non-load-bearing building enclosure type that transfer lateral environmental wind loads onto the primary structure, along with its own self-weight, but completely covers the building frame and shelters it from temperature swings, water (rain, snow, ice), wind, and generally separates the interior from the exterior environment. Like the cladding examples we saw earlier, a curtain wall ideally offers a continuous waterproof blanket wrapping the main building support. Unlike most of the previous non-load-bearing examples, the curtain wall can be completely independently framed, outboard of the primary building structure.

Glass architecture

The lineage of curtain wall construction in modern, Western, architectural history, traces back to the ideal of a crystalline modern architecture described in Scheerbart's *Glasarchitetktur*. Bruno Taut's glass pavilion (Figure 6.1), from the Cologne Werkbund Exhibition of 1914, proclaimed in the inscriptions on

▶ Figure 6.1

"The Glass House" Cologne, 1914. Architect: Bruno Taut, Berlin the pavilion that "Glass brings a new era" and "We feel sorry for the brick culture" because "without a glass palace, life becomes a burden". Taut's subsequent manifesto, produced with Walter Gropius, amongst other members of the Worker's Council for Arts, asserted the need to bring the crystalline geometry of the glass house and great architecture into the world of the broad masses.[1]

The prism of glass as a transparent glass skin has inspired architects since Bruno Taut's time, foreshadowing ever taller office buildings that could not be realized until the technology of material production became available for architectural construction some decades later. By relieving the exterior enclosure from supporting the weight of the building, and by careful coordination of building structure and systems, it has become possible to achieve the dream of the early twentieth century, defining a crystalline form by skimming in front of the structure entirely.

Materials and historical development

The materials and technology of light, non-weight-bearing, exterior enclosures derives historically, from glass and metal structures developed for temporary purposes – greenhouses to shelter plants and exhibit halls to shelter temporary displays and the visiting observers alike. As the industrial revolution rendered the materials used in these structures – glass and metal – to be inexpensive enough to be considered for everyday construction purposes, and as the idea of the newer materials and their expression as viable materials for use in architecture and construction evolved, these materials became increasingly accepted as a possible construction type for habitable, permanent buildings.

Early examples of greenhouse construction exploited the transparency and heat-retaining properties of glazed structures, now generally known as the greenhouse effect, to shelter exotic tropical plants like palm trees and citrus plants in colder northern climates. Royal collections of plants, reflecting the wealth of the owner as much as the latest curiosities collected or purchased from the newly discovered distant lands, if not an interest in science, evolved as a status symbol from the wealthiest families. The Palm House at Bicton (Figure 6.2), constructed c. 1843, by D.E. Bailey, Budleigh Salterton, Devon, UK, used glass and metal framing to create a shelter from a shimmering dome of many small pieces of glass held together with iron sash bars (see Figure 6.3). The design is attributed to John Claudius Loudon, whose *Remarks on the Construction of Hothouses*, 1817, and *Sketches of Curvilinear Hothouses*, 1818, included the introduction of the use of flat wrought iron bars with hook-shaped sections to support glass panels.[2]

◀ Figure 6.2

Palm House at Bicton, constructed c. 1843, by D.E. Bailey, Budleigh Salterton, Devon, UK. Exterior view

▲ Figure 6.3

Palm House at Bicton, constructed c. 1843, by D.E. Bailey, Budleigh Salterton, Devon, UK. Interior view

▲ Figure 6.4

Greenhouse, Royal Botanical Gardens, Kew, in 1844–1848, by architect Decimus Burton and ironworker Richard Turner

A similar structure was built at what is now known as the Royal Botanical Gardens, Kew, in 1844–1848, by architect Decimus Burton and ironworker Richard Turner, and incorporated wrought iron and blown glass to create a three-part structure (see Figure 6.4). The glass was supported on arches and frames tied together with pre-stressed cables, forming a 19-meter-high central space with 9 meter walkways on either side. Some accounts of the original glass described that it had been treated to achieve a light green film of copper deposit to improve shading effect yet still let in heat from the sun, which the greenhouse would then trap and use to warm the plants inside.

The Hallidie Building, 1904, constructed in San Francisco, designed by Willis Polk, is sometimes identified as the earliest North American example of the modern curtain wall. As shown in Figure 6.5, the rectangular grid of glass and framing skims in front of the floor slabs of the office building, maximizing the amount of daylight brought into the building. The reinforced concrete structure is pulled back from the glazing as the cantilevered slab edges present only a thin profile behind the glass. The transparency achieved by the simple framed grid of squares is adorned with cast iron filigree and a fire escape at either end, a requirement after the earthquake and fire of 1906. The glazing system on the Hallidie Building is clearly non-structural, spanning in front of several upper floors, outboard of the main structural system. Though the cast-iron edgings and cornice motifs hearken back to decorative interior millwork, these lace-like applied elements merely soften the otherwise utilitarian and rather modern glazing grid on the street facade.

At the Bauhaus building in Dessau, 1925–1926, Walter Gropius used a similar grid of light steel framing and glass but discarded all adornment (see Figures 6.6 and 6.7). The window system, taken from the vocabulary of utilitarian industrial structures and glazing for warehouses, also slides in front of the building structure, like the early curtain wall on the Hallidie Building. At the machine shops, the monumental glass walls borrowed from industrial factory windows and freed them from the building frame to achieve a modern, functional aesthetic with the promise of new materials and new production methods, representing a brighter, more socially responsible society.

While the evolution of the curtain wall construction type did not reach full maturity until after the Second World War, when aluminum production became widely available to the construction industry, it is worth examining the developments in construction in the earlier part of the century since they form the foundation for the systems which evolved in later years.

▶ Figure 6.5

Hallidie Building, 130 Sutter Street, San Francisco, CA, 1904, Willis Polk Architect, street view

▶ Figure 6.6

Bauhaus Dessau, Walter Gropius, exterior view showing extensive use of clear glass in steel industrial window framing

◀ Figure 6.7

Bauhaus Dessau, Walter Gropius, exterior view showing curtain wall skimming in front of building structure

Construction sequencing

Ever taller New York City office buildings proliferated in the years immediately prior to the crash of 1929 due to escalating property values and easy financing. The Chrysler Building and the Empire State Building, for example, vied for the claim of the tallest in the world. Construction of the Empire State Building, whose superstructure was erected within the span of only 11 months, 1930–1931, required ingenuity, expertise, and careful sequencing of construction.[3] Planning, procurement of materials, manufacture, and erection were all critical to meeting the incredibly ambitious construction schedule.

Photos from the construction of the Empire State Building (Figures 6.8–6.11) show that the sequenced installation of exterior enclosure was roughly

▶ Figure 6.8

Empire State Building construction photograph showing early installation of steel supports and stone cladding. No. 169, 6/25/30, 6th to 27th stories, 12th storey on stone

CHAPTER 6 Curtain walls

◀ Figure 6.9

Empire State Building construction photograph showing progression of construction on the primary building frame and the efficiency of subsequent installation of exterior enclosure. No. 221, 7/26/30, setting columns for 44th and 45th floors, 43rd floor of steel, 35th floor on metal trim, 29th floors on limestone, derricks on 6th floor setback setting limestone [on] 2nd, 3rd, 4th, 5th stories

▶ Figure 6.10

Empire State Building construction photograph showing sequence of construction. No. 254, 9/3/30, 72nd floor steel, metal trim 60th floor, floor arches 65th floor, stone 51st floor

◀ Figure 6.11

Empire State Building construction photograph showing nearly complete exterior enclosure. No. 367, 11/11/30, building of 1178′-9″ including 130′-5″ of tower (E-4 floor-97 stories) ten stories behind the erection of the primary steel building frame. This represents a key precursor to modern construction scheduling, which attempts to fireproof a steel frame and then enclose the exterior as quickly as possible, roughly also ten floors below the steel. This permits interior finish work to start as soon as the envelope is established. The steel structure was designed to incorporate a curtain wall support beam from which stainless steel strip mullions could be supported, outboard of the primary building structure. Limestone and aluminum veneer panels were quickly set within the stainless strips, backup masonry knee walls were constructed from inside the building, so windows could be set on the knee walls, spanning to the next spandrel above. Eliminating the scaffold and minimizing work from the exterior of the building expedited enclosure of the building, thanks to an integrated team of construction, design, and real estate entities using a fast track method of construction procurement.

Fireproofing

In addition to the evolution of building systems that would today be considered tangential to the architect's immediate responsibilities, developments in vertical transportation, or elevator design, and building systems to bring light, power, heat, fresh air, cooling, and running water that made it possible to build higher, the development of materials and systems of protection to building structures from fire is critical in the history of the evolution of tall buildings. As a rule, these concerns are less of a concern to the exterior skin where buildings are situated far from neighboring structures. However, in urban contexts, flame spread prevention between buildings, between tenants of a single building, and between floors becomes critical. Famous fires and disasters, such as the Great Fire of Chicago in 1871, spurred property owners to investigate noncombustible construction methods in the replacement buildings that were built afterwards.

William LeBaron Jenney, architect of the Home Insurance Building, one of the earliest skyscrapers constructed with an iron and steel structural frame, responded in the aftermath of the fire with a composite assembly of masonry, steel, and fired terracotta for fireproof structure, floor assemblies, and partitions. The Fair Store, for example, (Figure 6.12) incorporated the use of hollow terracotta planks and covers for the structural steel columns, girders, and beams. Though lighter steel decking has largely replaced the terracotta planking shown in the cutaway axonometric view, the concrete topping slab is similar to that used in modern high-rise construction below a floor finish. All surfaces of the steel were wrapped in noncombustible materials, as is common today, to shield the steel from direct exposure to the highest temperatures of a fire, and give inhabitants time to escape to safety.

Wrapping critical load-bearing structure with noncombustible materials is a procedure that today relies on testing to prove the ability of a given assembly to slow fire progression (see Figure 6.13). Underwriters' Laboratories (UL) and other standard testing agencies have established the number of minutes or hours that a given assembly can be reasonably expected to withstand a certain type and exposure to fire, allowing some level of protection and safety to

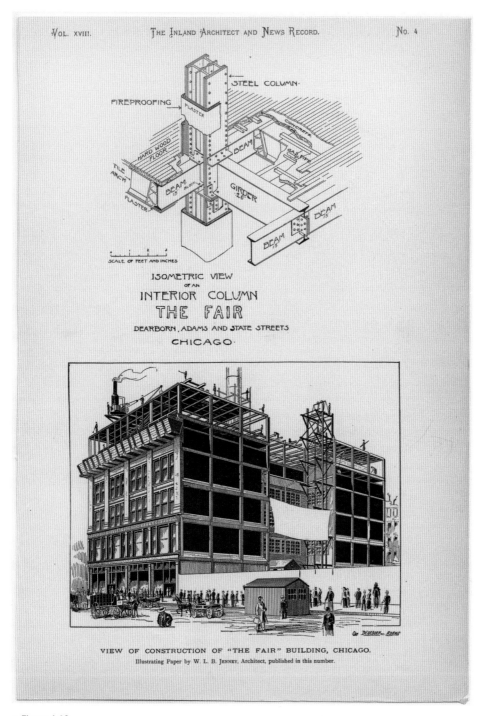

▲ Figure 6.12

William LeBaron Jenney, Fair Store, fireproofing primary building structure by wrapping with noncombustible materials such as terracotta tile

▶ Figure 6.13

Gypsum board wrapped steel, for fire protection. Noncombustible steel studs are used in conjunction with gypsum wallboard. Corner beads and screw heads are covered to achieve a clean surface, ready for paint

building inhabitants, allowing them to escape or rescuers to arrive. The impact on the exterior skin is often limited, especially for towers separated some distance from adjacent buildings. In dense urban contexts, many codes require some fire- and flame-spread treatment to limit the passage of smoke and flame between tenants on either side of a demising wall. For the exterior skin, it can be very important to prevent flame and smoke spread between floors. This can be addressed at the gap between the edge of structure and the interior face of the cladding.

Creation of a zone, roughly three feet or one meter in height, at each floor edge, is often sufficient to meet the code and limit fire spread in office buildings, even if they do not have full sprinkler coverage. The flame-spread prevention can also impact the architecture of the exterior enclosure since it can be as effective if it extends horizontally a comparable distance, three feet/one meter. These measures can allow sufficient time for occupants to escape before a fire jumps floors, or help prevent catastrophic failures of softened steel, illustrated by the post-fire photograph of Ernest Ransome's Pacific Coast Borax warehouse originally constructed in Bayonne, NJ (see Figure 6.14).

After World War II, Skidmore, Owings and Merrill's Lever House on Park Avenue in New York city represents an example of a proto-"stick-built" metal framing system with interesting tinted glass infill. Solex, a green body-tinted glass, was used in an early attempt to absorb the solar radiation hitting it during

◀ Figure 6.14

Fire damage to steel – Pacific Coast Borax Factory, Ernest Ransome Architect

the day. At night, the heat would dissipate equally to both the interior and exterior, but effectively 50 percent of that solar gain would be rejected back outdoors. In the case of the Lever House, the original framing members were actually fairly simple hot-rolled steel sections, channels, and angles, spanning vertically floor to floor, outboard of the slab edge and then in turn supporting horizontal members that held opaque spandrel glass hiding the building slab and perimeter steel from view. As shown in Figures 4.30 and 4.31, the project also used an early roof-mounted rig for installation of the exterior – a strategy commonly employed today for exterior building maintenance and window washing systems.

The sequence of construction used for the Seagram Building likewise incorporated expressed steel verticals at the exterior of the facade, using I-beam shaped members to stiffen and support the infill glazing (see Figures 6.15 and 6.16). At the corners, Mies van der Rohe carefully arranged the thin-edged flanges to emphasize the verticality of the skyscraper, in a very clear articulation of the support system. Thermal concerns make it difficult to justify these exposed radiator-like extensions of the primary building structure, in contemporary construction, unless thermal breaks or insulation are introduced, though they were expressive and elegant terminations of the building edge. The systems used in both the Lever House and the Seagram Building foreshadow the extruded aluminum systems identified today as "stick" systems.

The tube mullions that first came into wide use for curtain wall construction incorporated a clamping mechanism, in the form of an aluminum plate, to hold glass or other infill materials against the extruded aluminum profile. In combination, the mullion frame and infill materials create an enclosure system against environmental forces. While one-third the strength of steel, aluminum generally provides sufficient stiffness to resist wind loads while being approximately

▶ Figure 6.15

Seagram Building, New York City, Ludwig Mies van der Rohe, exterior view

▶ Figure 6.16

Seagram Building, New York City, Ludwig Mies van der Rohe, exterior view showing mullions articulated at the outside face with infill glazing

CHAPTER 6 Curtain walls

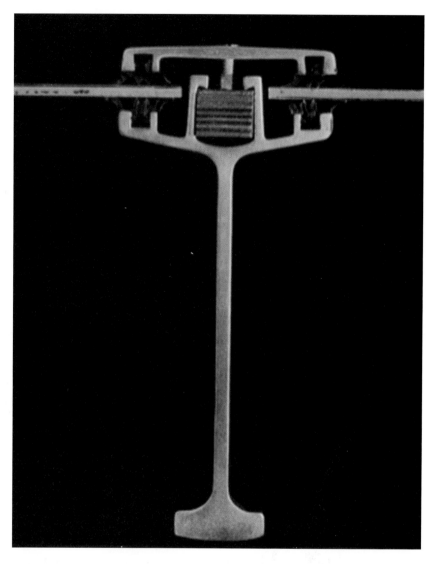

◀ Figure 6.17

Mullion from Jean Prouvé CIMT brochure cover

one-third the weight of steel. Many materials can be used to provide the facing material seen from the outside of the building. Provided the infill is sufficiently strong to resist the lateral forces from the environment, stone, metal, glass, and any other durable material that can withstand long-term exposure to the elements can be inserted in the channel, though it is usually called a "glazing" channel.

Early examples of stick-built mullion design using captured glazing evolved from systems like the Systeme Grille by the metal shop owned by the Ateliers Jean Prouvé (see Figure 6.17). The mullion in this type of assembly was profiled in the shape of a capital I, with flanges at the ends and minimal web, like a steel beam. This profile was, in some early prototypes, clad with glass, metal sheets, or an early form of insulated panel. The infill cladding was held in

◀ Figure 6.18

Pepsi-Cola Bldg, 500 Park Avenue, Natalie de Blois, Gordon Bunshaft, SOM, general exterior view

▼ Figure 6.19

Pepsi-Cola Bldg, 500 Park Avenue, Natalie de Blois, Gordon Bunshaft, SOM, view looking up at mullion

place, or captured, by the clamping action of a thin plate, continuously mirroring the length and weight of the mullion, applying pressure along the edges of the infill material by a series of screws that held the plate tight to the mullion. The frequently spaced screws created a positive mechanical attachment with many small fasteners distributing the forces from gusts of wind onto the mullion tube. Thus, a mechanism to hold the glazing in place was tied to a stronger backup member providing additional stiffness against lateral loads perpendicular to the face of the enclosure system.

The advantages of the simple I-beam shape lie principally in the structural efficiency of adding material at the extreme interior and exterior faces of the mullion, where it contributes to the strength in resisting loads. At the Pepsi-Cola Building, 500 Park Avenue, in New York City, Natalie de Blois, Gordon Bunshaft, and SOM implemented an elegant aluminum I-shaped mullion system with carefully proportioned, maximum size lights of clear glass, resulting in a transparent office building that floats against the original context of masonry apartment buildings that lined Park Avenue at the time the building was completed (see Figures 6.18 and 6.19). Mullion connection issues with flange-shaped mullions, however, are not insubstantial. Joining horizontal members to these profiles

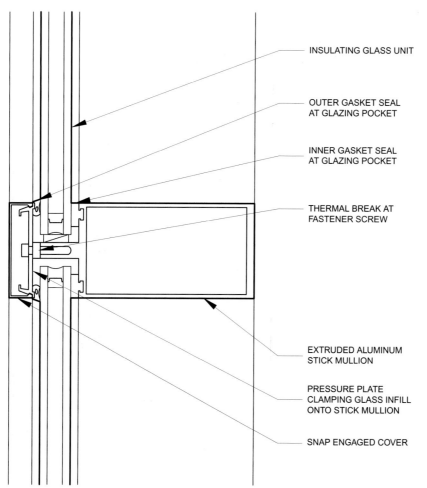

◀ Figure 6.20

Section detail at typical stick mullion profile

- INSULATING GLASS UNIT
- OUTER GASKET SEAL AT GLAZING POCKET
- INNER GASKET SEAL AT GLAZING POCKET
- THERMAL BREAK AT FASTENER SCREW
- EXTRUDED ALUMINUM STICK MULLION
- PRESSURE PLATE CLAMPING GLASS INFILL ONTO STICK MULLION
- SNAP ENGAGED COVER

requires multiple cuts during fabrication which, multiplied over the entire surface of a large building or skyscraper, can become cost prohibitive.

The most prevalent simple stick-built profile has been widely available with many minor proprietary variations for many decades (see Figure 6.20). This simple rectangular tube profile came into wide use after World War II, when the lengthy and costly mining, refining, and reducing of bauxite into aluminum shifted from war production for aircraft into the commercial and construction market (see Figure 6.21). From the basic tube, the mullion evolved to incorporate a tab with extruded ridges to accept the fastener screws of the pressure plate capturing the infill glazing, and indentations to receive extruded gaskets that isolate the infill material from the metal at both inside and outside, simultaneously forming both a vapor barrier and primary air barrier against the exterior environment. The tube at the back of the extrusion provides rigidity and stability, with its extreme flanges offering resistance to lateral loads at the tension and compression zones. The sides of the tube offer opportunities to make the connections of horizontal

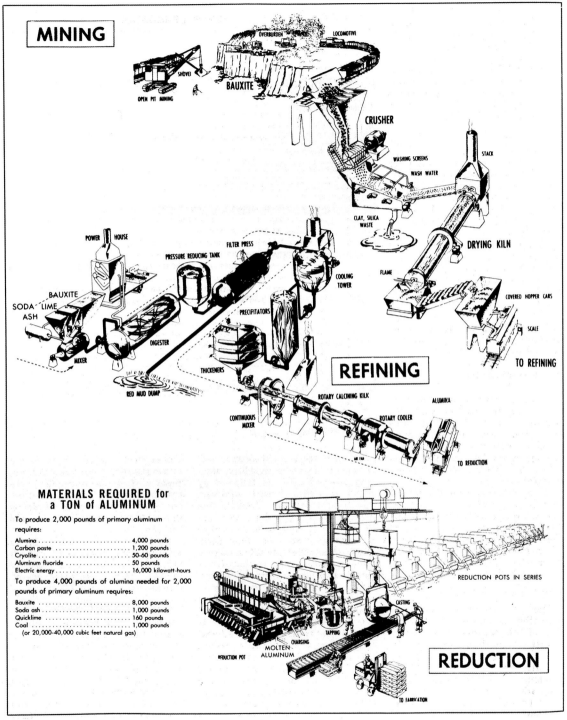

▲ Figure 6.21

How aluminum is mined, refined and reduced

mullions to vertical mullions using simple angles or specially profiled shear blocks. Alternatively, one side could be removed creating a C-shaped mullion, to minimize the number of fastener heads exposed to view, with access to the connector covered by thin, decorative, snap-engaged interior covers.

A whole host of projects executed using this mullion type have proliferated around the world, with further refinements such as the introduction of a rigid plastic thermal break, pad, or other nonconductive material. Provided that the nonconductive material is sufficiently stiff and strong as the aluminum, the filler material or other isolator introduces a solution that can be used in conjunction with insulating glass or other infill material to maintain a continuous, thermally broken line between heated interior spaces and cold outdoor climates (or vice versa). These extruded aluminum profiles are produced under pressure and heat, with semi-molten metal squeezed through die profiles, cut to lengths convenient for stocking, handling, and shipping. Standard 8" or 9" diameter dies are used most commonly for architectural purposes; though larger dies are possible they are rarely economical for construction projects, especially if the building size does not permit a significantly large order to be placed through the curtain wall supplier (see Figure 4.50). Extrusions are typically stocked in lengths to suit the majority of one- and two-storey installations, longer lengths being difficult to manipulate.

Most stick-built, tube-shaped mullion systems are attached to the edge of a building structure with two-part anchors. Like the masonry anchors we saw in Chapter 2, this split of the attachment into two pieces simplifies the method of adjusting so that inaccuracies in construction of the building structure are not transferred into the exterior skin system. It is rare for the position of structural steel and concrete to be accurate to within ½" of the target location, in the X-, Y-, and Z-axis. Curtain wall systems and, indeed, most modern exterior enclosures, are ideally not more than ⅛" from the desired final position, if not ¹⁄₁₆", therefore adjustable anchors with slots and simple mechanical fasteners are often ideal. Welded connections are possible, to minimize components, but are very highly variable, being most subject to inconsistencies in field conditions and workmanship under the more demanding construction site situations, where the building is not yet enclosed and all work must be done exposed to the elements. This contributes to the advantages of pre-engineered, mechanically attached anchorage types.

Height, movement, and the development of unitized systems

The stick mullion curtain walls we have reviewed thus far largely are constructed with vertical members spanning floor to floor, as the primary skin support (see Figure 6.22). Glass and other infill materials are usually sitting on horizontal members fixed to one vertical, and supported for gravity loads at the other end with some type of small slip joint to accommodate thermal expansion within the length of the horizontal member. Wind and other lateral forces are likewise transmitted from the horizontals to the verticals, then through the primary vertical mullion anchors onto the building frame.

▶ Figure 6.22

Typical stick mullion system, AAMA

1 Anchor
2 Stick
3 Horizontal
4 Spandrel panel
   (install from exterior)
5 Horizontal
6 Vision panel
   (install from interior)
7 Glazing step

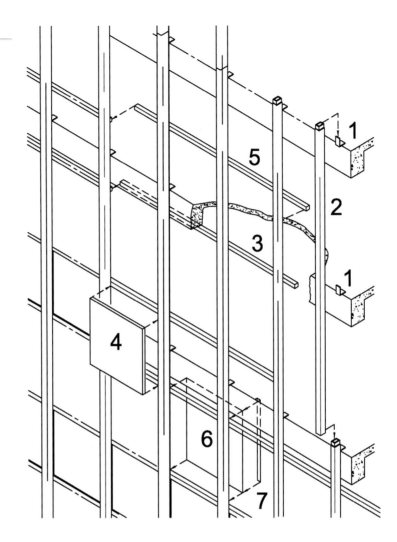

These systems are able to accept some sideways in-plane movements, whether from global building movements or seismic forces. The infill glazing, or other material, can be set with protection blocks at the edges, to limit rotation of the rectangular panel against the interior metal surface of the glazing pocket. The overall movements that can be accepted by such a system are, however, limited by the width of the glazing pocket and mullion. In general, the earliest examples of these systems were predicated on a subsequent scaffold drop of workmen with infill and glazing materials, applying the glazing stop pressure plate capturing the glass onto the mullions from the outside. Increasingly complicated mullions were developed to minimize the need for such additional, costly, suspended labor at the exterior of the building, either incorporating removable interior stops to the mullions or by limiting the use of these systems to low structures or areas of a building reachable from the ground.

▲ Figure 6.23

860–880 Lake Shore Drive, Chicago, IL, Ludwig Mies van der Rohe, construction view showing lifting of prefabricated "ladder" onto the building frame

Examples of hybrid stick systems include 860–880 Lake Shore Drive towers, Chicago, IL, designed by Ludwig Mies van der Rohe in 1949–1951 (Figure 6.23). These buildings were constructed, like the Seagram Building in New York City, using outboard steel mullions. From the construction image, in Figure 6.24, we see that ladder sections of the steel exterior skin framing were lifted into place. These ladders were prefabricated prior to lifting, to minimize labor on the floors. Likewise, infill panels were pre-glazed and set within the ladder assemblies. The project was sufficiently large, with the two towers combined, and sufficiently

▲ Figure 6.24

860–880 Lake Shore Drive, Chicago, IL, Ludwig Mies van der Rohe, construction view showing installation of infill panel into ladder framing

repetitive to be able to take advantage of economies of scale in the fabrication and production of the exterior enclosure.

With the increasing height of skyscrapers, in a never-ending competition to reach taller and taller heights, for reasons of national pride, commercial advertising advantages, or fame, a number of more sophisticated tools has been used to ensure structural soundness of the primary building frame as well as the skin enclosure surrounding it. The development of a better understanding of seismic behavior through increasingly sophisticated analysis and modeling tools in the later part of the twentieth century such as the rise of computers to handle repetitious mathematical work, much less the repetitive nature of the drafting efforts they required, permitted ever taller skyscrapers to be undertaken. Coupled with tuned mass dampers, some super tall buildings used sensors and large weights to "tune" the frequency of a tall building and offset building movements.

The more sophisticated understanding of large building movements and sway could not be accepted with simple stick mullion systems without significant penalty, cost or otherwise. Splitting the verticals into two parts created an

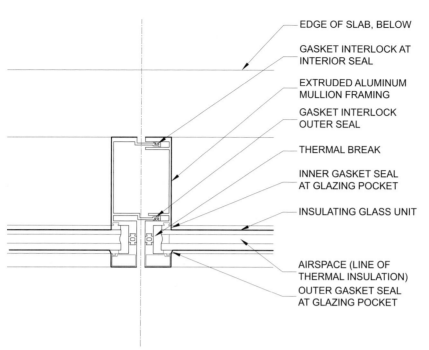

◀ Figure 6.25

Plan detail at typical split vertical mullion, for unitized construction

additional location for larger movements to be taken up, without danger to expensive and fragile glazing, or other infill materials. Using gaskets, nowadays typically neoprene, the man-made rubber substitute, dual seals could be maintained while permitting larger movements, even for tall buildings in seismically very active locations such as the Pacific Rim.

The two-part vertical mullions also suggested a completely different method of fabrication and installation of the curtain wall skin. By splitting the vertical into two halves, the connections of horizontal-to-vertical could easily be hidden from view by fastening through the open C-shaped half-mullion directly into the horizontal members (see Figure 6.25). If this is done at both sides of the horizontal, it is possible to create a ladder, which can be assembled off-site and lifted into place, in what could be called a unitized system. Glass and other infill materials can also be preinstalled on the ladders off-site, in controlled factories. Rather than shipping individual extrusions to the job site, this allows the field installation time to be minimized, and increases the complexity of glazing and assembly procedures in the shop, for maximum control and performance.

Expansion within the components of unitized curtain wall construction can be addressed by oversizing the slip joint at both horizontal and vertical split mullions. This is typically a fairly small dimension, but is cumulatively included in the joint movements that should be included and added to the structural deflections, seismic movements, fabrication, and erection tolerances to ensure the cleanest final installation. Tolerances on installation of curtain wall systems is on the order of a few millimeters, where tolerances on cruder building skeleton and slabs are significantly higher, so anchor design must be adjustable to accommodate these differences to ensure the smoothest and simplest site installation.

Unlike earlier precast concrete panelized examples we examined, molded from precast concrete to from a single element, with on-site infill of glass at openings, the framed ladder assemblies constructed from extruded aluminum are lightweight and become cost effective with larger-scale projects where prefabrication can simplify installation enough that significant customization becomes negligible to the construction budget, and in fact can save money. Lighter systems become reasonable to ship longer distances and eliminate the need for heavy cranes to lift and place the panels, and refined components reduce tolerance concerns considerably, within the curtain wall system itself. Especially when buildings are large enough to have substantial quantities of similar, repetitive units, like high-rise skyscrapers, customizations can become quite cost effective.

Since field work in full weather exposure is minimized, unitized curtain walls present a method of construction that can be used in areas where inclement conditions are prevalent at the anticipated time of installation. Where site labor is exceedingly expensive, it also minimizes the time, and therefore cost, required to enclose the building frame. Where the quality of site labor is unpredictable or poor, it maximizes the opportunity to produce a fully engineered, reliably water- and airtight assembly. In most unitized curtain walls, preset gaskets complete both seals without need for wet, sealant, installation. Thus there is no longer any need for additional access or work from the exterior. In more sophisticated systems, the anchors can be designed to adjust for anticipated construction and erection tolerances without the need for any additional exterior work. Often utilizing heavier extruded aluminum components, the anchor is also able to be designed to accept seismic movements in all directions, with very clever detailing.

Wind pressures and lateral forces

Like the introduction of architectural aluminum into the construction field in the post-World War II years, the development in understanding of wind behavior was borrowed from the aircraft industry, amongst others, and introduced into building design in the subsequent decades. The idea of modeling wind behavior physically, in a manner previously applied only to scientific and industrial endeavors, became interesting to the world of building construction in the competition to produce taller and taller skyscrapers. The sketch by structural engineer Leslie E. Robertson for the World Trade Center in New York City, designed by Minoru Yamasaki (Figure 6.26), shows the windward and leeward pressures on the simple, rectangular towers.

Looking back at the Earth from the moon, the ozone layer and cloud cover captures our atmosphere against the surface of the globe. As the Earth rotates on its daily cycle, relatively turbulence-free atmospheric winds are created. Closer in from this boundary layer, the effects of aerodynamic drag from the surface of the planet create localized conditions of mixing and turbulence, entirely the result of the local and regional geography. Immediate surroundings, whether open water, where winds can pick up speed, or highly packed, densely built urban conditions with many wind barriers, can affect wind behavior significantly. For most buildings, these wind conditions need to be evaluated for impact on both primary building structure and exterior enclosure.

Today, most building codes use data extrapolated from weather measurements of statistically maximum wind speeds (miles per hour or meters per second), which are then converted into pressures and suctions that could be seen on a building itself. These calculations tend to be less precise than wind tunnel studies, so if a customized analysis is performed, results may identify localized hot spots and unusual conditions in excess of minimum local code requirements.

The extrapolation formula used by the American Society for Civil Engineers, for example, calculates for factors such as geographical context conditions (open water, trees, and low suburban context, or urban areas with many structures, for instance), coastline distance (hurricane force gales), and incorporates assumptions regarding compound conditions. Where a gust of wind, typically measured in 3-second intervals, pushes on the wall of a building, the windward pressure can be evaluated. However, as that windward pressure blows over a building or around the corner, a leeward suction is created by the eddy of wind. Since wind is not typically a single-directional force, this leeward suction force on the other surface could occur at the same time that the surface was subject to wind suction forces as well. Thus the coefficients used in the calculation are used to develop a worst-case, relatively conservative pressure/suction force, typically expressed in pounds per square foot (PSF) or Newtons (N).

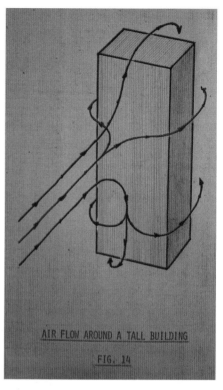

▲ Figure 6.26

Wind flow diagram developed by Leslie E. Robertson and Associates during design of the World Trade Center towers, New York City

Pressure maps like Figure 6.27, the diagram of the pressures and suctions anticipated on the World Trade Center Towers, can be developed using model codes, or more accurately, by physical testing. Using a wind tunnel, testing can be performed on tall buildings to model the impact of prevailing winds, urban context, and proposed building geometry. Building a scale model with pressure sensors inside all faces of the proposed building or buildings (see Figure 6.28) and inserting it into a scale model of the context allows the wind pressures and suctions to be simulated reasonably accurately. The combined context and building model are typically installed on a rotating platform so wind directions can be modeled three hundred sixty degrees, though the tunnel itself is set up with an aircraft engine generating airflow in only the direction of the length of the tunnel (see Figure 6.29). Contemporary wind tunnel testing generally produces wind contour maps for the surface of a specific building in its context that can be much more specific than model codes regarding peak hot spots of high pressures and, sometimes more onerously, peak suctions (see Figure 6.30).

The structural implication of localized wind loads like these pressure and suction maps can be significant for the skin. In addition, building sway and wind-induced movements can affect the size of joints between cladding panels.

▶ Figure 6.27

Pressure map diagram developed by Leslie E. Robertson and Associates during design of the World Trade Center towers, New York City

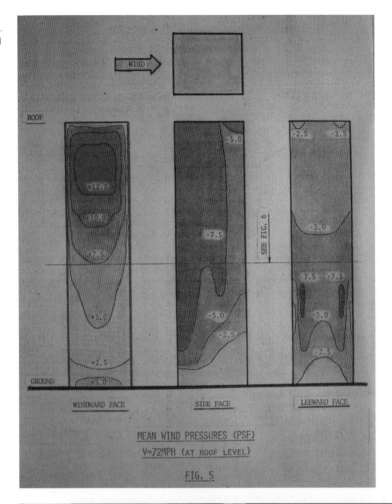

▶ Figure 6.28

WTC wind tunnel testing model, photograph of pressure differential sensing taps inside model of low podium buildings, at the bottom of the photo, in front of the taller WTC towers 1 and 2, at the top of the photo

▲ Figure 6.29

Photograph showing model of a building being tested in a wind tunnel with local context constructed in some detail on a circular platform so it can be rotated through 360 degrees of exposure

Movement in a three-dimensional structure from lateral forces can induce twist into the primary structure, and lead to inter-storey movement. Like seismic forces, these lateral movements can cause side-to-side movements in the plane of the exterior wall, perpendicular to the plane of the facade, and up/down. The cladding panels must be able to remain intact, for safety, for the duration of the environmental condition, for a statistically meaningful likelihood (50 year storm or 100 year storm, for example). The majority of building codes recognize and anticipate the need to address these conditions, and the design of cladding elements, panels, and joints between panels must take into consideration the effect of these loads.

Panelization

Panelized construction of exterior enclosures can be created from any number of materials and assemblies. The simplest, conceptually, is perhaps the precast concrete panel, which we reviewed in the cladding chapter when used in large format. Spanning column-to-column, the panel is isolated from localized slab deflections. However, in the case of a panel hung off the edge of slab, deflection between columns becomes significant to consider for the cladding system. Joints between panels must be sized wide enough to accommodate the component of the deflection to limit damage to the panels. The side-to-side movement, plus any limits to the joint filler, in an exposed concrete-to-concrete joint, for example, would have to be absorbed by the detail, or permanent damage to the enclosure panel could result.

Early examples of metal panels used as exterior skins include Jean Prouvé's Système Panneaux, a series of options for sandwich panels that used formed edges on the perimeter of the outer and inner facing panels to crimp them together. Variations on the panels were offered by his company in a number of standardized options, for instance with a small round porthole window, or with a

▶ Figure 6.30

Example of a pressure map developed from modern wind tunnel testing

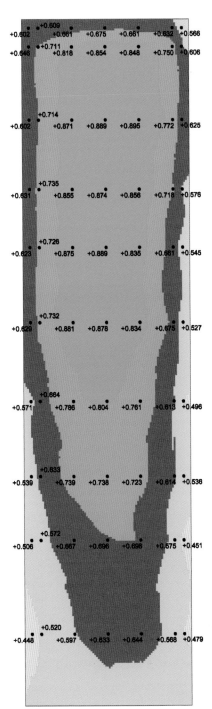

simple curved outer panel to provide stiffness without need for additional internal support framing, or stamped with three-dimensional profile for strength. A gasket at the perimeter was used to create a watertight single seal connection to the next panel, forming a stepped interlock to provide integral self-flashing. When front and back panels are used, they can sandwich insulating material between the exterior metal skin. The insulation can often also provide stiffening backup for thinner sheets, though studs and other framed supports may also be required. Commercial options today include foam-insulating fill. It is also common to find resin-core double-faced metal panels that provide improved stiffness with limited use of metal at surfaces, resulting in flatness and minimizing oilcanning.

The Alcoa Building in Pittsburgh, PA, from 1953, was designed by Harrison and Abramowitz to showcase the use of aluminum at every opportunity. Constructed to house the Aluminum Company of America (ALCOA), the 30-storey skyscraper was clad in ⅛" thick aluminum panels, stamped to provide additional rigidity to the skin (see Figure 6.31). The extensive use of aluminum

◀ Figure 6.31

Alcoa Headquarters, Pittsburgh, PA, Harrison & Abramowitz, 1953, cladding panels stamped to provide structural integrity against lateral forces

was intended to publicize the lightweight, modern nature of the material. By using aluminum throughout, the steel building frame could be significantly lighter and more efficient. This explains the prevalence throughout the present curtain wall industry of aluminum framing and unitized aluminum construction.

Pressure equalization and unitized mullion detailing

The concept of pressure equalizing the exterior skin design was touched upon in previous chapters, and is effectively a refinement of the dual-seal cavity wall, applied to increasingly sophisticated methods of detailing for exterior enclosure design. Whether by code analysis or using the results of a wind tunnel study, the pressure gradients across each building facade provide a map of the maximum pressure and suction forces that will need to be resisted by the cladding and the components supporting the cladding elements back to the primary structural frame. Overlaying the peak suction/pressure requirements over the elevations of a proposed design allow us to identify an optimal design pressure to use for sizing framing members and infill panel thicknesses. It may also be possible to identify areas where the cladding may require reinforcement or other special treatment to withstand the anticipated loads.

Within each zone of the skin assembly it is likely that natural separations between air cavities may already exist. In a tall building, convection forces may cause some natural rise in the air inside this cavity, drawing them upwards, unless there is some closure across the top, to prevent this stack effect within the cavity. Between horizontal flashing, typically once per floor in a brick or other rainscreen cavity wall example, there could effectively be a zone that will capture a volume of air. The volume of air in this chamber is essentially behind the first seal of a dual-seal assembly. Weep holes above the flashing in a cavity wall would provide the opportunity for the air inside the cavity to equalize in pressure to the atmospheric pressure outside the veneer, creating a pressure-equalized chamber. When wind gusts and localized suction pressures are applied to this zone of the building skin, the interior pressure within the pressure-equalized chamber will assist in preventing the infiltration of air (see Figure 2.12).

In curtain wall panels framed from extruded aluminum into a variety of customizable profiles, the mullions are profiled to hold the infill material in place. By definition, then, at least one cavity is created to fix the glass or other infill onto the picture frame of the support stiffeners. Within the mullion, the cavity can be used to create a pressure-equalized chamber. Using a second continuous, weathertight line of closure against the exterior and controlling the number of openings in the primary seal achieves this pressure equalization at the scale of the mullion. The weathertight enclosures can be achieved by selection of material and joint sealers that are naturally moisture- and wind-resistant. Wet seals incorporated into each of these systems, while perhaps more weathertight than dry gaskets, can also be more vulnerable to workmanship defects and may require scaffold access. At the height some skyscrapers are intended, scaffolding or other exterior access can be expensive, so double-gasket systems can be significantly more advantageous.

At the detail level, split mullion assemblies for unitized curtain walls can be profiled to prevent infiltration from environmental forces. Proprietary systems customized by each manufacturer have evolved to address these forces in a variety of ways.

To prevent water infiltration, mechanisms to allow sheeting of heavy rainfall without introducing any lip, stepped edge, or reverse slope are important. Misting rainfall or condensation of water vapor against the interior vapor barrier (secondary seal) can also present issues to be addressed – in some cases, mist may deposit unsightly atmospheric dust which can lead to corrosion, in the case of acid rain from pollution-laden regions, and drip patterns may affect the appearance over time. Control of condensation within the system, if conducted to the exterior, is typically the accepted, desirable solution. Details should be designed with the following responses to physical forces that may carry water into the system (see also Figure 2.13):

- Raising the interior of horizontal members in the chamber or sloping them toward the exterior becomes very useful
- Nesting the primary seal in a flush or slightly raised condition and covering or baffling any weep holes are also helpful
- Setting the secondary seal above and inboard of the primary seal contributes to this strategy of using gravity to naturally drain water infiltration or condensation to the exterior
- Baffling the inner surfaces with an upturned interior leg prevents wind-driven rain from bouncing up and into the chamber, and the return at the interior can be used to direct moisture outwards
- Spacing the gap between halves of the split mullion sufficiently wide apart to accommodate anticipated structural and building movements can also offer the opportunity to break any capillary action or inward movement from surface tension by creating a drip condition or cavity wide enough to minimize horizontal movement of water droplets.

Environmental forces

Water and the control of water vapor within the exterior wall assembly is one of the key elements that defines the modern curtain wall. The evolution from stacked monolithic components to cavity wall air gap design is overlaid by the principles of pressure equalization and rainscreen design at a detail level to achieve optimized performance under environmental forces. In addition to wind and lateral forces, the control of rain water and condensation is critical to the success of the system. Additionally, as extremes of temperature and region dictate, design detailing in temperate climates should also address snow and ice.

Snow which is allowed to collect on or near the vulnerable joints in a curtain wall system can be subject to thaw and re-freeze, compacting snow that may melt through heat lost from the interior over multiple daily cycles, forming an ice layer at the underside of the snow drift. When the water or compacted snow freezes again, it can, like in roofs, draw moisture up and under the primary

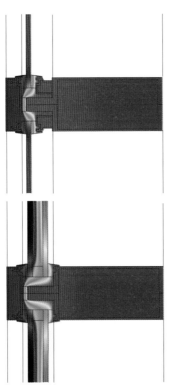

▲ Figure 6.32

Diagram of heat loss through the envelope with monolithic glass and stick mullion, no thermal break, versus insulating glass and mullion with thermal barrier minimizing heat flow

line of defense through capillary action of the ice. Local to a curtain wall mullion, this may be most vulnerable only at grade or roof intersections, if any. Any elements creating depth or special features providing a ledge for snow to settle can be subject to the same concerns, however, so a review of weather extremes and specific solutions to a given design may be advisable.

At the other extreme from snow and ice are sun and heat gain concerns. For both hot and cold outdoor conditions, the thermal break is the most effective solution to minimize heat gain or heat loss, respectively. By isolation of the elements exposed to the exterior temperature extremes, the interior portion of the system is protected from outside thermal swings from daily or yearly extremes. For infill materials incorporating significant capacity for insulation, the metal framing becomes the area most vulnerable to heat loss or heat gain (see Figure 6.32). In the case of conventional insulating glass used as infill in a framed system, the behavior of the glazing at the center of the light is typically the highest performance, since edge effects from the metal spacer used to set the dimension of the air cavity are negligible at the middle of the unit. Mullion framing and the performance of the system overall would be calculated with each of the components intended for use, and can be modeled with software prior to fabricating extrusions and assembling, when confirmation physical testing can be specified.

Examples of built projects incorporating innovative assemblies of interest include Jean Nouvel's design for the Institut du Monde Arabe, in Paris, from 1981, which captured diaphragm assemblies between two lights of glass in a unitized curtain wall at the south-facing elevation (see Figure 6.33). Expressing a modern interpretation of geometry and latticework found in Arab design, the glass and metal facade reflected the layering of the historic and cultural references from the Near East in a contemporary French context. Like a series of large and small camera lenses, the diaphragms were controlled automatically by light sensors and would adjust in response to daylight exposure to open or close and create shade in response to the environmental forces outside (see Figures 6.34 and 6.35).

The unitized curtain wall construction incorporated operable ventilators to provide access to the many intricate mechanisms from the interior (see Figure 6.36). Anchored to the edge of structure, the square units were divided into a large, central panel capturing a sub-grid of framed glass sandwiching metal diaphragms. The central panels were arranged around a large central diaphragm with rings of smaller- and medium-sized lens diaphragms at the perimeter of the panel. This early attempt at creating an active, responsive smart wall appears no longer to be functioning, a problem common to all new, intricate technology, especially in light of the long-term durations expected from architectural

◀ Figure 6.33

Jean Nouvel, Institut du Monde Arabe, Paris, overall exterior view

◀ Figure 6.34

Jean Nouvel, Institut du Monde Arabe, Paris, detail view from exterior showing shade mechanisms which borrowed from camera lens technology and were connected to daylight sensors to filter sunlight

▶ Figure 6.35

Jean Nouvel, Institut du Monde Arabe, Paris, detail view from interior showing shade mechanism

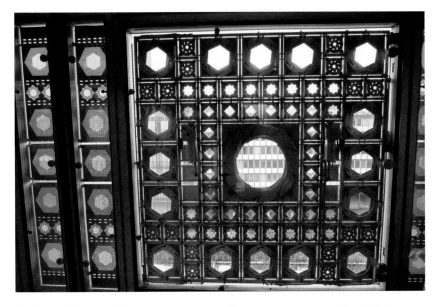

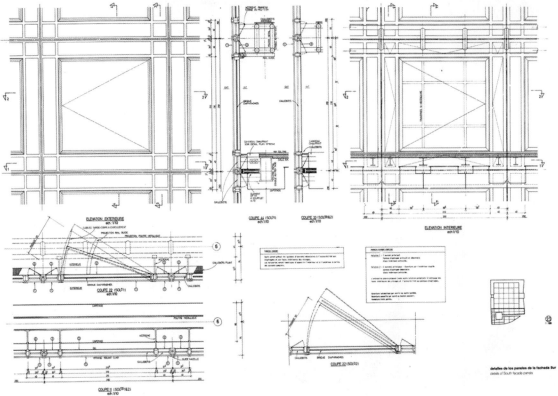

▲ Figure 6.36

Jean Nouvel, Institut du Monde Arabe, Paris, drawing showing unitized construction and operable interior windows for access to sun shade diaphragm mechanisms

construction. Even in the fixed position, the intensity of light at the interior is modulated by the filigree of the decorative lens shade sandwiched in the units of the Institut du Monde Arabe.

New technologies have always interested architects for their design potential and have often been adopted reasonably rapidly into the building construction and architectural vocabulary, though the life expectancy of most buildings often exceeds the capacity of a newer, smarter technology, especially in the initial stages of its development. Like the operable diaphragms in the Institut du Monde Arabe, the possibility of capturing any number of assemblies and products within a unitized framing system begins to achieve the prefabricated Taylorized assembly process for architecture and construction that the industrial revolution brought to production of consumer goods. Where sufficient repetition of regular, repeatable units is possible, significant customizations can become economically advantageous to implement. As the development of active systems for environmental controls evolves in coming years, and as energy costs rise throughout the world, options for mechanizing the exterior skin of buildings may become increasingly viable.

Finishes and form

The traditional finish of extruded aluminum was anodization, a controlled process to produce a natural oxidation layer that inhibits excess future degradation of the outer surface of aluminum, even when exposed to rain and moisture or UV radiation from the sun. Over time, the anodization can still pit and deteriorate, at which point very little can be done to salvage the structural integrity of the aluminum, and replacement may become the only option. The development of UV-resistant paint systems that can be bound to the aluminum at a molecular level, initially through proprietary Kynar and Hylar finishes, for example, created surface treatments that offered protection from sun exposure as well as a moisture barrier between the base aluminum substrate and the outside environment. Though these finishes can be somewhat vulnerable to abrasion from touch, the thin coatings, often electrostatically applied in the shop, effectively increase the durability of metal to prevent oxidation and can usually be touched up if scratched, if not repainted in the field where necessary. Metallic paints can be significantly more expensive than nonmetallic, due to the process of creating a reflective appearance, often with mica flecks.

Material selection naturally affects how a skin assembly behaves when exposed to solar radiation – both absorbtance and thermal mass, for instance, should be considered. In addition, color can have a significant effect on heat gain. Light surfaces have a higher albedo (the reflectivity of a surface against heat gain) than darker surfaces of the same material, with the same absorbance. By coupling a heavy, massive masonry wall with a black painted exterior face and a layer of glazing on top, the Trombe Wall in the Kelbaugh House, for instance, creates sufficient heat to supply the one-room-deep house with nearly all the warmth required throughout the winter in the relatively cold New Jersey climate (see Figures 2.9 and 2.10). The south facing proto-double wall is oriented to

▶ Figure 6.37

Sauerbruch Hutton GSW Tower, Berlin, Germany – overall view

capture maximum exposure to any available sunlight, and the convection current within the cavity causes heat to rise in the two-storey chamber. By carefully locating air inlets at the bottom and vents at the top of the poured concrete wall, the heat circulates into the occupied spaces on the shaded north side of the house.

At the GSW Tower, Berlin, Germany, Sauerbruch Hutton designed one of the first tall buildings to be constructed, in 1999, after reunification as headquarters for the German real estate company GSW Immobilien AG. A low, three-storey building at the ground is topped by a tall, thin tower, rising 22 stories in the neighborhood around Checkpoint Charlie (see Figure 6.37). The transparent facades of the tower are constructed with colorful vertical louvers at the West

▲ Figure 6.38

Sauerbruch Hutton GSW Tower, Berlin, Germany, detail view of facade – colored sunshade

elevation (Figure 6.38), captured within a double wall assembly that can be vented in hot summer months to draw heat away from the primary inner envelope (see Figure 6.39). Brightly colored in random-patterned, lighter hues of red when the west louver blades are shut, the building changes dramatically in appearance when occupants elect to open the blinds for the view. At the east facade, where heat loads present a lesser concern, the curtain wall incorporated silver louvered shade panels outboard of the primary enclosure system, providing a thinner double construction assembly (see Figure 6.40). The limited depth of the building permits natural ventilation through operable windows in the relatively temperate summers to offer sufficient air flow and circulation across the building and reduce cooling requirements. An air foil at the top of the west wall, above the double wall, uses prevailing winds to generate wind movements across the top of the double wall cavity and draw additional air upwards through the cavity (see Figure 6.41). In conjunction with daylighting opportunities from both sides of the tower, these features result in an environmentally responsible and efficient union of skin and architecture.

▶ Figure 6.39
Sauerbruch Hutton GSW Tower, building section showing airfoil above west wall

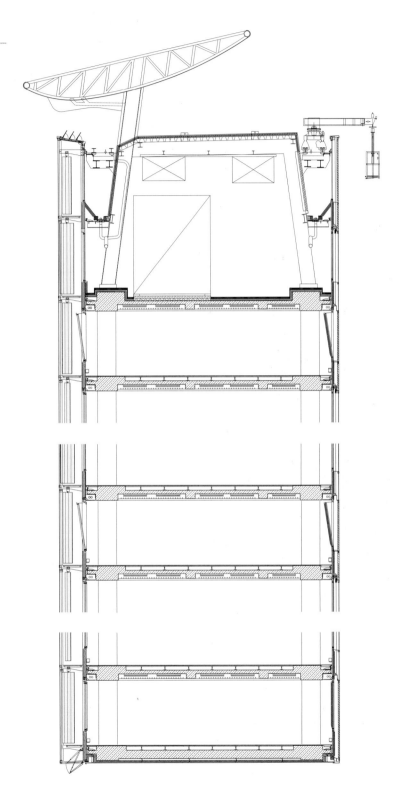

◀ Figure 6.40

Sauerbruch Hutton GSW Berlin Skyscraper, detail of facade photo – louvered side

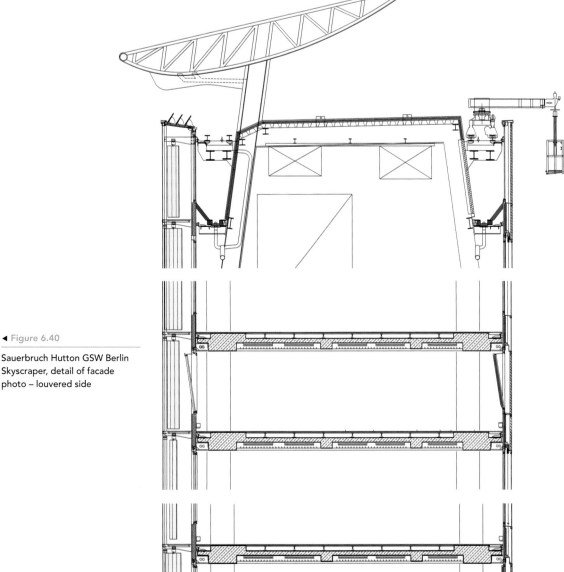

▲ Figure 6.41

Sauerbruch Hutton GSW – detailed wall section showing deep double wall at west wall and thinner louvered double wall at east

CHAPTER 6 Curtain walls   237

Unitized curtain wall

The ability to customize extruded aluminum profiles for larger projects has allowed the design of exterior building enclosures for tall buildings to be modified and customized for aesthetic intent and cost savings. With sufficiently large projects, enough repetition can be achieved to make modifications to basic profiles quite advantageous. The unitized construction of a facade (Figure 6.42) can appear fairly similar to the stick-built curtain wall system, although a split mullion system is typically used (see Figure 6.43). To accommodate deflections and movements, the split mullions must incorporate a slip joint. This can mean a slightly wider face profile, especially for the split horizontal, which is likely to see live load deflections from the larger structure, in addition to local thermal expansion and panel-to-panel movements.

Attachment to the primary structure can be quite similar to the anchoring of a stick, though the unit is typically prefabricated off-site into a ladder truss with horizontal

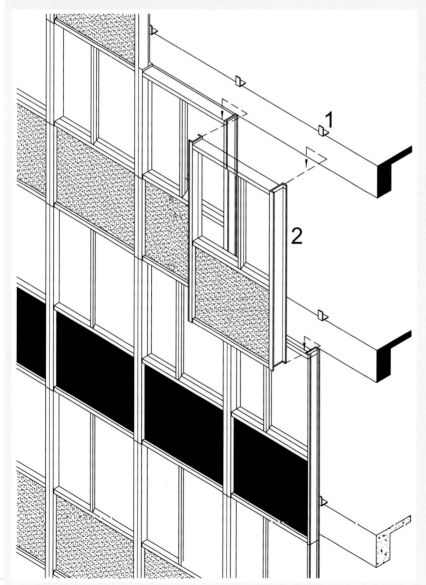

◀ Figure 6.42

Unitized curtain wall diagram
1 Anchor
2 Unitized curtain wall panel

Unitized curtain wall *continued*

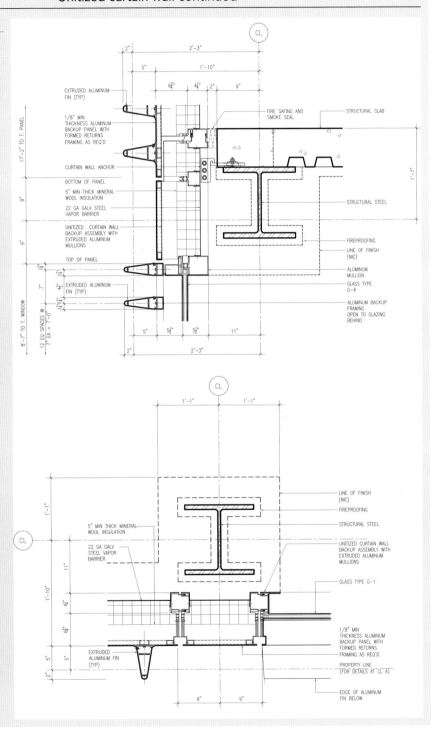

▶ Figure 6.43

Section at stack joint and plan detail at curtain wall, Columbia University Northwest Corner Building, Rafael Moneo/DBB

Unitized curtain wall *continued*

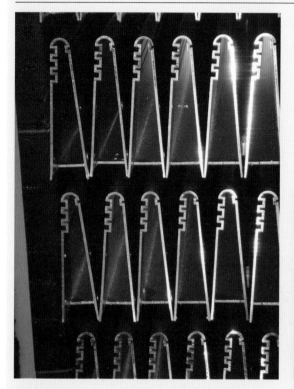

▲ Figure 6.44

Clear anodized extruded aluminum louvers, fabricated and ready to assemble

at the northwest corner of the main Columbia University Campus, a rainscreen panel was mounted to the outside face of a unitized curtain wall panel. To achieve the desired corrugated aluminum surface with deep sunshades over the vision areas, a series of extruded aluminum profiles was attached to a backing sheet, supported on a system of tube framing. These panels were fabricated by a very qualified metal working company, out of state, and shipped to the project site in New York. Extrusions were cut to length (Figure 6.44), pre-drilled, and sheet material was cut to the desired shapes and profiles (Figure 6.45). The material was assembled in the comfort of a work-

▲ Figure 6.45

End caps fabricated and pre-drilled

extrusions to support any change in material and provide rigidity to the assembly. Adjustment for inaccuracies in the construction of the crude building frame must be leveled out by the anchor assembly, therefore two components are typically used. The attachment of the first mechanical fastener onto the structure is the first point where adjustment can be made to true out the rough primary structure. The second connection point, between the first anchor component and the second, can adjust in a second axis. Slotted holes can be used to provide load transfer in one direction, while allowing adjustment in another direction. The final attachment point of the anchor onto the unit provides the third axis of adjustment. Thus, construction tolerances in the ½"–1" range (12mm–25mm) in the raw structure can be accommodated in the final curtain wall installation, which can be expected to be erected within 1⁄16" to 1⁄8" (3–6mm) accuracy.

In the fabrication of the curtain wall of the interdisciplinary science building designed by Rafael Moneo, with Davis Brody Bond architect of record, completed in 2011,

▲ Figure 6.46

Assembly of rainscreen panels

Unitized curtain wall *continued*

▲ Figure 6.47

Connections for rainscreen panel components

▲ Figure 6.48

Assembly of rainscreen panels in progress

▲ Figure 6.49

Rainscreen panels – fabrication complete, crated and ready to ship to the construction site

▲ Figure 6.50

Unitized backup panels in production on conveyor belt

shop (Figure 6.46), with framing and connectors fastened together (Figure 6.47). Extruded aluminum fin profiles and backing sheets were fastened to the outer face of the support framing (Figure 6.48). These materials formed the outer skin of the final assembly, and eventually would become the primary water- and rain-screening layer. They were crated (Figure 6.49) and shipped by truck, lifted onto the vacant floors, ready to be connected to the curtain wall backing units.

The inner curtain wall unit framing was assembled at a curtain wall fabrication shop in another city. Using a propri-etary unitized mullion extrusion system, framing elements, vision glass, and insulation were fabricated and assembled on a conveyor belt (Figure 6.50). In a clean, well-lit, interior workspace, the critical insulation and watertight vapor barrier seals were installed in each unit. These units were also crated and shipped to the job site, after completion (Figure 6.51). Off-site production such as this can be advantageous for projects where site labor is expensive, provided the project incorporates sufficient repetitive elements.

At the field, uncrated rainscreen panels were attached to the backup curtain wall units (Figure 6.52). The

Unitized curtain wall *continued*

▲ Figure 6.51

Completed backup unitized panels waiting to be crated and shipped to the site

▲ Figure 6.53

Lifting curtain wall unit into place

▲ Figure 6.52

Rainscreen panels mated to curtain wall backup units on-site

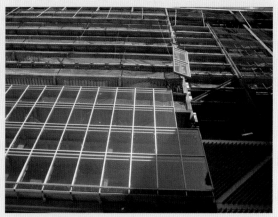

▲ Figure 6.54

Lifting glazed curtain wall unit into place

combination of rainscreen panels on curtain wall units together created a single entity. The units were pushed to the edge of the building on wheeled trolleys and then lifted into place (Figures 6.53 and 6.54). As shown in Figures 6.55 and 6.56, the exterior enclosure in this project incorporated full height glass at office areas, and metal panels with diagonal fin louvers where cross bracing structure occurred and opacity was desirable for programmatic purposes. At lab areas (Figures 6.57 and 6.58), sunshades over vision areas and diagonal fins at cross bracing bays syncopate the Broadway and 120th Street facades.

Unitized curtain wall *continued*

▲ Figure 6.55
View of offices at campus elevation, floor-to-ceiling glass

▲ Figure 6.56
View of interface between all glass offices and fin louvered panels

▲ Figure 6.57
Broadway elevation, view looking up at fin louvered panels and windows with sunshades

Unitized curtain wall *continued*

▲ Figure 6.58
120th Street elevation, looking up at café glass and fin louvered panels above

Notes

Chapter 1

1. For a discussion of the controversy regarding Portland cement, see Cecil Elliott, *Technics and Architecture*, Cambridge: MIT Press, 1992, pp. 150–157.

Chapter 2

1. Säynatsälo town hall was officially opened in 1952. Aalto won the competition of 1949. City of Jyväskylä, "Brief History of Säynatsälo." http://www.jyvaskyla.fi/saynatsalo/english/briefhistory

Chapter 3

1. "From the 1790s onwards, however, several patents were registered for nail-making machines and in 1807 Jesse Reed invented machine that could produce 60,000 good-quality nails in a day." Colin Davies, *The Prefabricated Home*, London: Reaktion Books Ltd, 2005, p. 46.
2. Food and Agriculture Organization of the United Nations 2011 Report, Global Forest Resources Assessment and U.S. Department of Agriculture, Forest Service, "U.S. Forest Facts and Historical Trends", FS-696–M, September 2001, Forest Inventory and Analysis Program, http://fia.fs.fed.us/library/briefings-summaries-overviews/docs/ForestFactsMetric.pdf

Chapter 4

1. Design Forum Finland. "Auli Suortti-Vuorio, Kaj Franck (1911–1989) Professor, Designer." http://www.designforum.fi/kaj_franck_en.

Chapter 5

1. Kenneth Frampton, *Modern Architecture: A Critical History*, London: Thames and Hudson, Third Edition, 1980 rev 1992, pp. 82–83.

Chapter 6

1. Kenneth Frampton, *Modern Architecture: A Critical History*, London: Thames and Hudson Ltd., 1992, Chapter 13, pp. 116 and 117.
2. George Kohlmaier and Barna von Sartory. *Houses of Glass: A Nineteenth Century Building Type*. Originally published by Prestel-Verlag, Munich 1981; Cambridge, MA; London, England: The MIT Press, 1986, page 139.
3. Carol Willis, *Building the Empire State: A Rediscovered 1930's Notebook Charts the Construction of the Empire State Building.* New York: W. W. Norton & Company in association with The Skyscraper Museum, 1998.

List of figure credits

1.1 Ricardo Liberato, Wikimedia Commons
1.2 Løken, Wikimedia Commons
1.3 Photo by Bryan Feldman, www.greatwallforum.com
1.4 Carl Krebs
1.5 Construction of Walls and Arches, Vaults and Domes from The History of Architecture by Banister Fletcher, edited by Dan Cruikshank, 20th Edition, p. 198; Architectural Press, 1996. Reproduced by permission of Taylor & Francis Group; and with kind permission of the RIBA Library Books & Periodicals Collections.
1.6 © Corky Buczyk, courtesy of Shutterstock
1.7 Library of Congress LC-DIG-ppmsc-06601, Detroit Publishing Co., 1905
1.8 GrouchoMarxist, Wikimedia Commons License
1.9 RaBoe, Wikimedia Commons License
1.10 Roger Ulrich, by permission
1.11 Metsavend Wikimedia Creative Commons license
1.12 Apfel51 Wikimedia Creative Commons license
1.13 Faqscl, Wikimedia Creative Commons License
1.14 Faqscl, Wikimedia Creative Commons License
1.15 Patrickringgneberg, Wikimedia Creative Commons License
1.16 Photo by Monfie, stitching by Colin, Wikimedia Creative Commons License
1.17 Sara L. MacKenzie
1.18 Sara L. MacKenzie
1.19 Sara L. MacKenzie
1.20 R. Mark, Princeton University
1.21 Allard Schmidt, Wikimedia Creative Commons license
1.22 Colegota, Wikimedia Creative Commons License
1.23 Mayine L. Yu
1.24 Mayine L. Yu
1.25 Mayine L. Yu
1.26 Photo by Chicago History Museum/UIG via Getty Images
1.27 Jessica Sheridan, Wikimedia Creative Commons license
1.28 American building: materials and techniques from the first colonial settlements to the present by Condit, Carl W. Copyright 1968 Reproduced with permission of University of Chicago Press – Books in the format Other Book via Copyright Clearance Center.
1.29 Plan for garden wall, Thomas Jefferson Architectural Drawings, 1817–1819. Albert and Shirley Small Special Collections Library, University of Virginia
1.30 Karen Blaha, Attribution-ShareAlike 2.0 Generic (CC BY-SA 2.0)
1.31 Arq. Juan Andrés Fernández, www.jaf.com.uy
1.32 Dieste & Montañez S.A.
1.33 Dieste & Montañez S.A.
1.34 Dieste & Montañez S.A.
1.35 Dieste & Montañez S.A.
1.36 Arq. Juan Andrés Fernández, www.jaf.com.uy

1.37	Arq. Juan Andrés Fernández, www.jaf.com.uy
1.38	Based on Figure 12, Olin, Harold B., Construction: Principles, Materials and Methods, 8th Edition, New York: John Wiley & Sons, 2006. Redrawn by Mary McConnell
1.39	Mayine L. Yu
1.40	Mayine L. Yu
1.41	© Alvar Aalto Museum
1.42	Nico Saieh
1.43	Hassan Bagheri
1.44	Louis I. Kahn Collection, The University of Pennsylvania and the Pennsylvania Historical and Museum Commission, photo by John Ebstel. By permission of Keith De Lellis Gallery
1.45	Louis I. Kahn Collection, The University of Pennsylvania and the Pennsylvania Historical and Museum Commission, photo by John Ebstel. By permission of Keith De Lellis Gallery
1.46	Glen-Gery Corporation
1.47	Gilbreth, Frank B. *Bricklaying System.* New York and Chicago: The Myron C. Clark Publishing Co., 1909
1.48	André Wolf Pause / Mayine L. Yu
1.49	Glen-Gery Corporation
1.50	André Wolf Pause / Mayine L. Yu
1.51	Based on Chin, Ian. "A Thin-stone Veneer Primer." *Architectural Record.* Volume 178, No. 7 (June 1990)
1.52	Mayine L. Yu
1.53	Cornell University Library
1.54	Ford, Edward R., The Details of Modern Architecture, Volume 1, Austin Hall, p. 28, © 1990 Massachusetts Institute of Technology, by permission of The MIT Press.
1.55	Austen Clark
1.56	Ada Karmi-Melamede
1.57	Ada Karmi-Melamede
1.58	© Corky Buczyk, courtesy of Shutterstock
1.59	Sarah Ackerman
1.60	Sarah Ackerman
1.61	Sarah Ackerman
1.62	Courtesy of Kevin Roche John Dinkeloo and Associates LLC
1.63	Mayine L. Yu, © 2012 Artists Rights Society (ARS), New York / ADAGP, Paris
1.64	Immanuel Nicholas Iyadurai
1.65	Avery Architectural and Fine Arts Library, Columbia
1.66	Archives, Avery Library, Columbia University
1.67	Archives, Avery Library, Columbia University
1.68	Archives, Avery Library, Columbia University
1.69	Immanguel Giel Wikimedia Creative Commons licenses
1.70	Immanguel Giel Wikimedia Creative Commons licenses
1.71	Mayine L. Yu
1.72	Mayine L. Yu
2.1	Based on Flynn, John E., Segil, Arthur W., and Steffy, Gary R. *Architectural Interior Systems: Lighting, Acoustics, Air Conditioning.* New York: Van Nostrand Reinhold, 1988
2.2	Library of Congress, Prints & Photographs Division, LC-DIG-ggbain-15108

2.3   Library of Congress, Prints and Photographs Division, Gottscho-Schleisner Collection, LC-DIG-gsc-5a29821 DLC
2.4   Ford, Edward R., The Details of Modern Architecture, Volume 1, Morgan Library, p. 56, © 1990 Massachusetts Institute of Technology, by permission of The MIT Press
2.5   Based on Flynn, John E., Segil, Arthur W., and Steffy, Gary R. *Architectural Interior Systems: Lighting, Acoustics, Air Conditioning.* New York: Van Nostrand Reinhold, 1988
2.6   Based on Olin, Harold B., Construction: Principles, Materials and Methods, 8th Edition, New York: John Wiley & Sons, 2006. Moisture control, pp. 104–10
2.7   Based on Meyers, James. "Building Façade Watertightness." *Progressive Architecture*. Vol. 72, No.11 (November, 1991). Redrawn by Mary McConnell
2.8   Meyers, James. "Building Façade Watertightness." *Progressive Architecture*. Vol. 72, No.11 (November, 1991). Redrawn by Mary McConnell
2.9   Mayine L. Yu
2.10 Mayine L. Yu
2.11 Stainaker, Judith J. and Harris, Ernest C. Structural Design in Wood. New York: Van Nostrand Reinhold, 1989. Redrawn by Mary McConnell
2.12 Reprinted, with permission, from the American Architectural Manufacturers Association
2.13 Reprinted, with permission, from the American Architectural Manufacturers Association
2.14 Scott Gilchrist / Archivision Inc
2.15 Detail: Erblin Bucaliu
2.16 Ford, Edward R., The Details of Modern Architecture, Volume 2: 1928 to 1988, Exeter, p. 326, © 1996 Massachusetts Institute of Technology, by permission of The MIT Press
2.17 Zache, Wikimedia Commons Multi-license with GFDL and Creative Commons CC-BY-SA-2.5 and older versions (2.0 and 1.0)
2.18 Hisao Suzuki
2.19 Hisao Suzuki

3.1   Frontispiece, Marc-Antoine Laugier, *Essai Sur l'Architecture*, London: printed for Stanley Crowder and Henry Woodgate, 1756
3.2   Library of Congress, Prints and Photographs Division, Historic American Buildings Survey or Historic American Engineering Record, Reproduction Number HABS NC,82-KER.V,1-1
3.3   Library of Congress, Prints and Photographs Division, Historic American Buildings Survey or Historic American Engineering Record, Reproduction Number: HAER WYO,1-KESTO.V,1-B-1
3.4   Mayine L. Yu
3.5   Library of Congress, Prints and Photographs Division, Historic American Buildings Survey or Historic American Engineering Record, Reproduction Number HABS TENN,32-NOET,1A--20
3.6   Drawing from *Architectural Graphic Standards*, Tenth Edition, John Ray Hoke, Jr. FAIA Editor in Chief, Wiley, 2000. This material is reproduced with permission of John Wiley & Sons, Inc.
3.7   With kind permission of the RIBA Library Books & Periodicals Collections
3.8   Library of Congress, Prints and Photographs Division, Historic American Buildings Survey or Historic American Engineering Record, Reproduction Number HABS NEB,54-NIOB,11-4

3.9 Drawing from *Architectural Graphic Standards*, Tenth Edition, John Ray Hoke, Jr. FAIA Editor in Chief, Wiley, 2000. This material is reproduced with permission of John Wiley & Sons, Inc.
3.10 Library of Congress, Prints & Photographs Division, FSA-OWI Collection, LC-USF342-T01-008270-A DLC
3.11 Library of Congress, Prints and Photographs Division, Historic American Buildings Survey or Historic American Engineering Record, Reproduction Number HABS RI,3-NEWP,44-1
3.12 Drawing from *Architectural Graphic Standards*, Tenth Edition, John Ray Hoke, Jr. FAIA Editor in Chief, Wiley, 2000. This material is reproduced with permission of John Wiley & Sons, Inc.
3.13 Photograph Copyright 1987 Wayne Cable (.com)
3.14 Yusuke Oono
3.15 Ford, Edward R., The Details of Modern Architecture, Volume 1, Jacobs House, p. 330, © 1990 Massachusetts Institute of Technology, by permission of The MIT Press
3.16 Levittown Public Library
3.17 Levittown Public Library
3.18 Levittown Public Library
3.19 Levittown Public Library
3.20 Library of Congress, Prints and Photographs Division, Historic American Buildings Survey or Historic American Engineering Record, Reproduction Number HABS MASS,9-LIN,16--12
3.21 Library of Congress, Prints and Photographs Division, Historic American Buildings Survey or Historic American Engineering Record, Reproduction Number HABS MASS,9-LIN,16--16
3.22 Library of Congress, Prints and Photographs Division, Historic American Buildings Survey or Historic American Engineering Record, Reproduction Number HABS MASS,9-LIN,16--5
3.23 Mayine L. Yu
3.24 Mayine L. Yu
3.25 Mayine L. Yu
3.26 Mayine L. Yu
3.27 Mayine L. Yu
3.28 Mayine L. Yu
3.29 Mayine L. Yu
3.30 Mayine L. Yu
3.31 Mayine L. Yu
3.32 Mayine L. Yu
3.33 Mayine L. Yu
3.34 Mayine L. Yu
3.35 Mayine L. Yu
3.36 Christina Richards, Dreamstime
3.37 Mayine L. Yu
3.38 Mayine L. Yu
3.39 Areta Pawlynsky
3.40 Areta Pawlynsky
3.41 © Donald Corner & Jenny Young/Artifice Images
3.42 Hisao Suzuki
3.43 Hisao Suzuki
3.44 Whit Slemmons

3.45 Whit Slemmons
3.46 Whit Slemmons
3.47 Mayine L. Yu
3.48 Mayine L. Yu
3.49 Mayine L. Yu
3.50 © Greg Murphey, courtesy of Studio Gang Architects
3.51 © and courtesty of Studio Gang Architects
3.52 © Greg Murphey, courtesy of Studio Gang Architects
3.54 Areta Pawlynsky
3.53 © Greg Murphey, courtesy of Studio Gang Architects

4.1   Sara L. Mackenzie
4.2   Sara L. Mackenzie
4.3   Sara L. Mackenzie
4.4   Liang Ssu-ch'eng. Fairbank, Wilma, Ed. *Chinese Architecture: A Pictorial History*. Dover Publications, 2005
4.5   Liang Ssu-ch'eng. Fairbank, Wilma, Ed. *Chinese Architecture: A Pictorial History*. Dover Publications, 2005
4.6   Raphael Azevedo Franca, Wikimedia Commons
4.7   Sam D'Cruz/Shutterstock
4.8   Bernd and Hilla Becher, *Framework House*, Sonnabend Gallery, New York
4.9   Library of Congress, Prints and Photographs Division, Historic American Buildings Survey or Historic American Engineering Record, Reproduction Number HABS PA,67-YORK,12-6
4.10 Mayine L. Yu
4.11 Architectural Press Archive / RIBA Library Photographs Collection
4.12 J. Russell and Sons
4.13 Plate XVII Design for storefronts, *Badger's Ilustrated Catalague of Cast Iron Architecture*
4.14 Mayine L. Yu
4.15 Marco 2000 licensed under the Creative Commons Attribution-Share Alike 2.0 Generic license
4.16 Library of Congress, Prints and Photographs Division, Historic American Buildings Survey or Historic American Engineering Record, Reproduction Number HABS ILL,47-PLAN.V,1-6
4.17 Library of Congress, Prints and Photographs Division, Historic American Buildings Survey or Historic American Engineering Record, Reproduction Number HABS ILL,47-PLAN.V,1-9
4.18 Mayine L. Yu
4.19 Steve Hall © Hedrich Blessing, courtesy of Studio Gang Architects
4.20 © and courtesty of Studio Gang Architects
4.21 Diderot Encyclopedie – Glaces Soufles, plate 13, 14, 36
4.22 W. Cooper Crown Glass
4.23 Mario Savoia
4.24 David K. Shea / Mayine L. Yu
4.25 Crown Copyright
4.26 Mayine L. Yu
4.27 unclepepin
4.28 Olin, Harold B., *Construction: Principles, Materials and Methods*, 8th Edition, New York: John Wiley & Sons, 2006. This material is reproduced with permission of John Wiley & Sons, Inc.

4.29	Hisao Suzuki
4.30	Mayine L. Yu
4.31	Avery Architectural and Fine Arts Library, Columbia University
4.32	Symbiosus, Claudia and Christian von Montfort, Creative Commons license
4.33	Mayine L. Yu
4.34	Anthony Sieverding
4.35	Anthony Sieverding
4.36	Mayine L. Yu
4.37	Ernest L. Ransome, Alexis Saurbrey, *Reinforced Concrete Buildings*, 1912
4.38	Mayine L. Yu
4.39	Bruno Taut *Modern Architecture.* London: The Studio Limited, 1929
4.40	Mayine L. Yu
4.41	Mayine L. Yu
4.42	Mayine L. Yu
4.43	Mayine L. Yu, © 2012 Artists Rights Society (ARS), New York / ADAGP, Paris
4.44	Mayine L. Yu, © 2012 Artists Rights Society (ARS), New York / ADAGP, Paris
4.45	Ford, Edward R., *The Details of Modern Architecture*, Volume 1, Swiss Pavilion, p. 252, © 1990 Massachusetts Institute of Technology, by permission of The MIT Press
4.46	Nick Haynes
4.47	Margherita Spiluttini
4.48	© 2012, Herzog & de Meuron Basel
4.49	Reprinted, with permission, from the American Architectural Manufacturers Association
4.50	Olin, Harold B., *Construction: Principles, Materials and Methods*, 8th Edition, New York: John Wiley & Sons, 2006. This material is reproduced with permission of John Wiley & Sons, Inc
4.51	Reprinted, with permission, from the American Architectural Manufacturers Association
4.52	Nancy Peterson, Azon
4.53	Raymond McGrath
4.54a	Centre Pompidou – MNAM – Kandinsky Library
4.54b	© 2012 Artists Rights Society (ARS), New York / ADAGP, Paris
4.54c	© 2012 Artists Rights Society (ARS), New York / ADAGP, Paris
4.54d	© 2012 Artists Rights Society (ARS), New York / ADAGP, Paris
4.55	Smallbones, Wikimedia Commons license
4.56	Erik Christensen, Wikimedia Commons license
4.57	Seier + Seier, Wikimedia Commons license
5.1	Mayine L. Yu / Mary McConnell
5.2	Wagner: Werk – Museum Postsparkasse, Vienna
5.3	Werkmeister, Wikimedia Commons license
5.4	Gryffindor, Wikimedia Commons license
5.5	Gryffindor, Wikimedia Commons license
5.6	Gryffindor, Wikimedia Commons license
5.7	Foster & Associates
5.8	Foster & Associates
5.9	Foster & Associates
5.10	Tim Hursley
5.11	Tim Hursley
5.12	A. Holford Smith

5.13   A. Holford Smith
5.14   Mayine L. Yu
5.15   Sidney Freedman
5.16   Mayine L. Yu
5.17   Mayine L. Yu
5.18   Copyright 2012 Benny Chan / Fotoworks
5.19   Courtesy of Neil M. Denari Architects
5.20   Copyright 2012 Benny Chan / Fotoworks
5.21   Copyright 2012 Benny Chan / Fotoworks
5.22   Copyright 2012 Benny Chan / Fotoworks
5.23   Reprinted, with permission, from the American Architectural Manufacturers Association
5.24   Macrae / "Starrett-Lehigh Building, New York City," Architectural Record 71 (Jan 1932)
5.25   "Starrett-Lehigh Building, New York City," Architectural Forum 55 (Oct 1931), p. 491
5.26   "Starrett-Lehigh Building, New York City," Architectural Forum 55 (Oct 1931), p. 490
5.27   Macrae / "Starrett-Lehigh Building, New York City," Architectural Record 71 (Jan 1932), p. 32
5.28   "Starrett-Lehigh Building, New York City," Architectural Record 71 (Jan 1932), p. 31
5.29   Library of Congress, Prints and Photographs Division, Historic American Buildings Survey or Historic American Engineering Record, Reproduction Number HABS WIS,51-RACI,5-2
5.30   Library of Congress, Prints and Photographs Division, Historic American Buildings Survey or Historic American Engineering Record, Reproduction Number HABS WIS,51-RACI,5-22
5.31   Library of Congress, Prints and Photographs Division, Historic American Buildings Survey or Historic American Engineering Record, Reproduction Number HABS WIS,51-RACI,5-15
5.32   Library of Congress, Prints and Photographs Division, Historic American Buildings Survey or Historic American Engineering Record, Reproduction Number HABS WIS,51-RACI,5-7
5.33   Reprinted, with permission, from the American Architectural Manufacturers Association
5.34   Mayine L. Yu
5.35   Mayine L. Yu
5.36   Foster + Partners
5.37   Foster + Partners
5.38   Foster + Partners
5.39   Duccio Malagamba
5.40   Rafael Moneo
5.41   Rafael Moneo
5.42   Duccio Malagamba

6.1   Bruno Taut *Modern Architecture*. London: The Studio Limited, 1929
6.2   Mark A C Photos
6.3   Helga Ruddy
6.4   Newton and Co
6.5   Library of Congress, Prints and Photographs Division, Historic American Buildings

	Survey or Historic American Engineering Record, Reproduction Number HABS CAL,38-SANFRA,149--1
6.6	Charlotte Nordahl, Wikimedia Commons license
6.7	Cethegus, Wikimedia Commons license
6.8	The Erection Views of Empire State Building; A Photograph Album Compiled by Starrett Brothers and Eken Incorporated, 2002.05: 49; 271. The Skyscraper Museum
6.9	The Erection Views of Empire State Building; A Photograph Album Compiled by Starrett Brothers and Eken Incorporated, 2002.05: 49; 272. The Skyscraper Museum
6.10	The Erection Views of Empire State Building; A Photograph Album Compiled by Starrett Brothers and Eken Incorporated, 2002.05: 71; 377. The Skyscraper Museum
6.11	The Erection Views of Empire State Building; A Photograph Album Compiled by Starrett Brothers and Eken Incorporated, 2002.05: 78; 416. The Skyscraper Museum
6.12	Fair Store, Chicago, IL, 1892, 1896. Jenney & Mundie, architects. Archival Image Collection, Ryerson and Burnham Archives, The Art Institute of Chicago. Digital File # IA1804_0822 © The Art Institute of Chicago
6.13	United States Gypsum
6.14	The Atlas Portland Cement Company, *Reinforced Concrete in Factory Construction*, New York, NY: The Atlas Portland Cement Company, 1907. Fig. 109
6.15	Mayine L. Yu
6.16	Mayine L. Yu
6.17	© 2012 Artists Rights Society (ARS), New York / ADAGP, Paris
6.18	Mayine L. Yu
6.19	Mayine L. Yu
6.20	Mayine L. Yu
6.21	Olin, Harold B., *Construction: Principles, Materials and Methods*, 8th Edition, New York: John Wiley & Sons, 2006. This material is reproduced with permission of John Wiley & Sons, Inc
6.22	Reprinted, with permission, from the American Architectural Manufacturers Association
6.23	Photograph by Hedrich Blessing, Chicago History Museum, Negative # HB-13809-J3
6.24	Photograph by Hedrich Blessing, Chicago History Museum, Negative # HB-13809-Y4
6.25	Davis Brody Bond LLP
6.26	The Records of Leslie E Robertson and Associates (LERA): World Trade Center Construction Images, 2004.02.2502, ca. 1969–1973; The Skyscraper Museum
6.27	The Records of Leslie E Robertson and Associates (LERA): World Trade Center Construction Images, 2004.02.2489, ca. 1969–1973; The Skyscraper Museum
6.28	The Records of Leslie E Robertson and Associates (LERA): World Trade Center Construction Images, 2004.02.2293, ca. 1969–1973; The Skyscraper Museum
6.29	RWDI
6.30	RWDI
6.31	Library of Congress, Prints and Photographs Division, Historic American Buildings Survey or Historic American Engineering Record, Reproduction Number HABS PA-6724-1
6.32	David K. Shea / Mayine L. Yu
6.33	Hisao Suzuki

6.34	Sara L. Mackenzie
6.35	Mayine L. Yu
6.36	Ateliers Jean Nouvel
6.37	Carlos Prieto/AGE Fotostock/Getty Images
6.38	Klara Kristina
6.39	Sauerbruch Hutton
6.40	© Annette Kisling
6.41	Sauerbruch Hutton
6.42	Reprinted, with permission, from the American Architectural Manufacturers Association
6.43	Davis Brody Bond LLP
6.44	Davis Brody Bond LLP
6.45	Davis Brody Bond LLP
6.46	Davis Brody Bond LLP
6.47	Davis Brody Bond LLP
6.48	Davis Brody Bond LLP
6.49	Davis Brody Bond LLP
6.50	Davis Brody Bond LLP
6.51	Davis Brody Bond LLP
6.52	Davis Brody Bond LLP
6.53	Davis Brody Bond LLP
6.54	Davis Brody Bond LLP
6.55	Davis Brody Bond LLP
6.56	Davis Brody Bond LLP
6.57	Davis Brody Bond LLP
6.58	Davis Brody Bond LLP

Bibliography

"4,000 Houses per Year: Levitt & Sons Demonstrate the Economies of the Bid Building Management, Purchasing, and Production Know-How." *Architectural Forum* 90 (Apr 1949): 84–92.

Allen, Edward. *Fundamentals of Building Construction*, 3rd edn. New York: John Wiley & Sons, Inc., 1999.

Ambrose, James. *Building Construction: Enclosure Systems*. New York: Van Nostrand Reinhold, 1990.

American Architectural Manufacturers' Association, *Aluminum Curtain Wall Design*, Des Plaines, IL, 1979.

Anderson, Stanford. *Eladio Dieste: Innovation in Structural Art*. New York: Princeton Architectural Press, 2004.

Archieri, Jean-François. *Prouvé Cours du CNAM 1957–1970: Essai de reconstitution du cours à partir des archives Jean Prouvé*. Liège: Pierre Mardaga, éditeur, 1990.

Atlas Portland Cement Company. *Reinforced Concrete in Factory Construction*. New York, NY: The Atlas Portland Cement Company, 1907.

Badger, Daniel D. *Illustrations of Iron Architecture, Made by the Architectural Iron Works of the City of New York*. New York: Baker & Godwin, 1865.

Badger, Daniel D. *Badger's Illustrated Catalogue of Cast Iron Architecture (The Architectural Ironworks of the City of New York)*. With a new introduction by Margot Gayle. New York: Dover, 1981.

Banham, Reyner. *A Concrete Atlantis: U.S. Industrial Building and European Modern Architecture, 1900–1925*. Cambridge, Mass: MIT Press, 1986.

Banham, Reyner. *Architecture of the Well-Tempered Environment*. London: Architectural Press, 1984.

Bradshaw, Vaughn. *Building Control Systems*, 2nd edn. New York: John Wiley & Sons, Inc., 1993.

Brand, Ronald E. *Architectural Details for Insulated Buildings*. New York: Van Nostrand Reinhold, 1990.

Brandford Landau, Sarah and Condit, Carl. *The Rise of the New York Skyscraper, 1865–1913*. New Haven and London: Yale University Press, 1996.

Brock, Linda. *Designing the Exterior Wall: An Architectural Guide to the Vertical Envelope*. New York: John Wiley & Sons, Inc., 2005.

Brookes, Alan and Grech, Chris. *The Building Envelope and Connections*. Oxford: Architecture Press, 1996.

Brownlee, David B. and De Long, David G. *Louis Kahn*. Los Angeles: The Museum of Contemporary Art, 1991.

Campbell, James W. P. and Pryce, Will. *Brick: A World History*. London: Thames & Hudson Ltd, 2003.

Chin, Ian. "A Thin-stone Veneer Primer." *Architectural Record*, 178, 7 (June 1990): 108–113.

Ching, Francis D. K. and Adams, Cassandra. *Building Construction Illustrated, Third Edition*. New York: John Wiley & Sons, Inc., 2001.

Compagno, Andrea. *Intelligent Glass Facades: Material, Practice, Design*. Basel, Boston: Birkhäuser-Verlag, 2002.

Condit, Carl. *American Building Art: The Nineteenth Century*. New York: Oxford University Press, 1960.
Condit, Carl. *The Chicago School of Architecture: A History of Commercial and Public Building in the Chicago Area, 1875–1925*. Chicago: University of Chicago Press, 1964.
Condit, Carl. *American Building: Materials and Techniques from the First Colonial Settlements to the Present*. Chicago: University of Chicago Press, 1968.
Cooper, William. *Crown Glass Cutter and Glazier's Manual*. Edinburgh: Oliver & Boyd, 1835.
Dal Co, Francesco. *Kevin Roche*. New York: Rizzoli, 1985.
Davies, Colin. *The Prefabricated Home*. London: Reaktion Books Ltd, 2005.
Diderot, Denis and d'Alembert. *Encyclopédie ou Dictionnaire Raisonné des Sciences, des Arts, et des Métiers*. Livourne: Imprimerie des editeurs, 1770–1775.
Elliot, Cecil. *Technics and Architecture*, Cambridge: MIT Press, 1992.
Fletcher, Sir Banister (ed. Dan Cruickshank). *A History of Architecture, Twentieth Edition*. Oxford: Architectural Press, 1996.
Flynn, John E., Segil, Arthur W. and Steffy, Gary R. *Architectural Interior Systems: Lighting, Acoustics, Air Conditioning*. New York: Van Nostrand Reinhold, 1988.
Food and Agriculture Organization of the United Nations, "Global Forest Resources Assessment. FRA 2010." Rome: Food and Agriculture Organization of the United Nations, 2010.
Ford, Edward. *The Details of Modern Architecture, Volume 1*. Cambridge: M.I.T. Press, 1990.
Ford, Edward. *The Details of Modern Architecture, Volume 2: 1928 to 1988*. Cambridge: M.I.T. Press, 1996.
Frampton, Kenneth. *Modern Architecture: A Critical History*, 3rd edn. London: Thames and Hudson Ltd, 1992.
Friedman, Donald. *Historical Building Construction: Design, Materials & Technology*. New York, London: W.W. Norton & Company, 1995.
Galerie Jousse Seguin–Galerie Enrico Navarra; Prat, Nathalie. *Jean Prouvé*. Paris: Galerie Jousse Seguin, Galerie Enrico Navarra, 1998.
Garlock, Maria E. Moreyra and Billington, David P. *Félix Candela: Engineer, Builder, Structural Artist*. Princeton, NJ: Princeton University Art Museum; New Haven: Yale University Press, 2008.
Gayle, Margot and Gayle, Carol. *Cast-Iron Architecture in America: The Significance of James Bogardus*. New York: Norton and Co., 1998.
Giedion, Sigfried. *Space, Time and Architecture: The Growth of a New Tradition*. Cambridge, Mass: Harvard University Press, 1971.
Gilbreth, Frank B. *Bricklaying System*. New York and Chicago: The Myron C. Clark Publishing Co., 1909.
Guiness, Desmond and Sadler, Jr., Julius Trousdale. *Mr. Jefferson, Architect*. New York: The Viking Press, 1973.
Guise, David. *Design and Technology in Architecture*. New York: Van Nostrand Reinhold, 1991.
Herzog, Thomas. *Façade Construction Manual*, Basel, Boston: Birkhauser-Publishers for Architecture, 2004.
Herzog & de Meuron. "Complete Works: 137 Dominus Winery, Yountville, California, USA; Project 1995, realization 1996–1998." http://www.herzogdemeuron.com/index/projects/complete-works/126–150/137-dominus-winery.html
Hildebrand, Grant. *Designing for Industry: the Architecture of Albert Kahn*. Cambridge: MIT Press, 1974.
Hix, John. *The Glasshouse*. London: Phaidon Press Limited, 1996.

Hoke, John Ray, Jr., Editor-in-Chief. *Ramsey/Sleeper Architectural Graphic Standards*, 10th edn. New York: John Wiley & Sons, Inc., 2000.

Huber, Benedikt and Steinegger, Jean-Claude (eds), Lieven, Alexander (trans). *Jean Prouvé: Prefabrication, Structures and Elements*. London: Pall Mall Press, 1971.

Huxtable, Ada Louise. *Pier Luigi Nervi*. New York: G. Braziller, 1960.

Jett, Megan. "Supreme Court Building in Jerusalem/Ada Karmi-Melamede Architects & Ram Karmi." *ArchDaily*, 21 November 2011. Accessed 08 September 2012. http://www.archdaily.com/184543.

Kaltenbach, Frank (ed.). *Detail Practice: Translucent Material: Glass, Synthetic Materials, Metal*. Birkhäuser: Edition Detail, 2004.

Karmi-Melamede, Ada. "The Supreme Court Building, Israel." *Perspecta*, 26 (1990): 83–96.

Kohlmaier, Georg and von Sartory, Barna. *Houses of Glass: A Nineteenth Century Building Type*. Originally published by Prestel-Verlag, Munich 1981 (Cambridge, MA; London, England: The MIT Press, 1986).

Krinsky, Carol Herselle. *Gordon Bunshaft of Skidmore, Owings & Merrill*. New York, NY: Architectural History Foundation; Cambridge, Mass: MIT Press, 1988.

Lechner, Norbert. *Heating, Cooling, Lighting: Design Methods for Architects*. New York: Wiley, 2000.

Lehman, Gary Paul. "Inventors and Inventions: The 2 × 4." *Forest History Today*, 23; reprinted from *Chronicle, Early American Industries Association*, 47 (September 1994): 90–91. http://www.foresthistory.org/publications/FHT/FHT1995/Lehmann.pdf

Levene, Richard C. and Márquez Cecilia, Fernando (eds). *Jean Nouvel 1987–1998*. Madrid: El Croquis Editorial, 1998.

Liang, Ssu-ch'eng (ed. Wilma Fairbank). *A Pictorial History of Chinese Architecture: a Study of the Development of its Structural System and the Evolution of its Types*. Cambridge: MIT Press, 1984.

Liang, Ssu-ch'eng (ed. Wilma Fairbank). *A Pictorial History of Chinese Architecture: A Study of the Development of its Structural System and the Evolution of its Types*. Mineola, NY: Dover Publications Inc., 2005.

Liu, Laurence G. *Chinese Architecture*. London: Academy Editions, 1989.

Lohan, Dirk. *Global Architecture Detail: Mies van der Rohe, Farnsworth House, Plano, Illinois, 1945–1950*. Tokyo: A.D.A. Edita, 1976.

Loughran, Patrick. *Falling Glass: Problems and Solutions in Contemporary Architecture*. Boston, MA: Birkhäuser-Publishers for Architecture, 2003.

McGrath, Raymond and Frost, A.C. *Glass in Architecture and Decoration* (new edn rev. Raymond McGrath). London: The Architectural Press, 1961.

Mark, Robert. *Experiments in Gothic Structure*. Cambridge, Mass: MIT Press, 1982.

Márquez Cecilia, Fernando and Levene, Richard C. *Herzog & de Meuron, 1981–2000: The Nature of the Artifice*. Madrid: El Croquis Editorial, 2000.

Márquez Cecilia, Fernando and Levene, Richard C. *Herzog & de Meuron, 1998–2002: The Nature of the Artifice*. Madrid: El Croquis Editorial, 2002.

Minor, Joseph. "Selecting Glass for Tall Buildings." *Architectural Record*, 178, 7 (June 1990): 105–107.

Myers, James C. and Ruggiero, Stephen S. "Building Façade Watertightness." *Progressive Architecture*, 72, 11 (November, 1991): 113–119.

Nichols, Frederick Doveton. *Thomas Jefferson's Architectural Drawings: Compiled and with Commentary and a Check List*. Boston: Massachusetts Historical Society; Charlottesville: Thomas Jefferson Memorial Foundation and The University Press of Virginia, 1984.

O'Connor, Thomas. "Exterior Wall Sealant Joint Design." *Progressive Architecture*, 72, 11 (November, 1991): 121–125.

Oesterle, Eberhard. *Double-Skin Facades: Integrated Planning*. Munich, London: Prestel, 2001.

Olgyay, Viktor. *Design with Climate: Bioclimatic Approach to Architectural Regionalism*. Princeton: Princeton University Press, 1963.

Olgyay, Aladar and Olgyay, Victor. *Solar Control and Shading Devices*, Princeton University Press, 1957.

Olin, Harold B., Schmidt, John L., and Lewis, Walter H. *Construction: Principles, Materials & Methods*, 5th edn. Chicago, Ill: Institute of Financial Education; Danville, Ill: Interstate Printers and Publishers, 1983.

Persily, Andrew. "Technics: Envelope Design Guidelines for Federal Office Buildings." *Progressive Architecture*, 72, 11 (March, 1992): 39–45.

Ransome, Ernest and Saurbrey, Alexis. *Reinforced Concrete Buildings: A Treatise on the History, Patents, Design, and Erection of the Principal Parts Entering into a Modern Reinforced Concrete Building*. New York: McGraw-Hill Book Co., 1912.

Rice, Peter and Dutton, Hugh. *Structural Glass*. London: E & FN Spon, 1995.

Rush, Richard D. (ed.). *The Building Systems Integration Handbook*. Boston, London: Butterworth-Heineman, 1986.

Sands, Herman. *Wall Systems: Analysis by Detail*. New York: McGraw-Hill, 1986.

Schittich, Christian (ed.). *Glasbau Atlas/Glass Construction Manual*. Basel; Boston: Birkhäuser, 1998.

Schittich, Christian (ed.). *Building Skins: Concepts, Layers, Materials*. München: Edition Detail; Basel: Birkhäuser, 2001.

Schulitz, Helmut C., Sobek, Werner, and Habermann, Karl J. *Steel Construction Manual*. Boston: Birkhäuser, 2000.

Simmons, H. Leslie. *Olin's Construction: Principles, Materials, and Methods*, 8th edition. Hoboken, New Jersey: John Wiley & Sons, Inc., 2011.

Stainaker, Judith J. and Harris, Ernest C. *Structural Design in Wood*. New York: Van Nostrand Reinhold, 1989.

"Starrett-Lehigh Building, New York City." *Architectural Forum*, 55 (Oct 1931): 483–492.

"Starrett-Lehigh Building, New York City." *Architectural Record,* 71 (Jan 1932): 30–35.

Taut, Bruno. *Modern Architecture*. London: The Studio Limited, 1929.

U.S. Department of Agriculture, Forest Service, "U.S. Forest Facts and Historical Trends." FS-696–M, September 2001, Forest Inventory and Analysis Program, http://fia.fs.fed.us/library/briefings-summaries-overviews/docs/ForestFactsMetric.pdf

Ursprung, Philip (ed.). *Herzog & de Meuron: Natural History*. Montréal, Québec: Canadian Centre for Architecture; Baden, Switzerland: Lars Müller, 2002.

Viskichil, Larry A. and Dean, Grant Talbot. *Chicago at the Turn of the Century in Photographs: 122 Historic Views from the Collections of the Chicago Historical Society*. New York: Dover Publications, 1984.

Wiggington, Michael. *Glass in Architecture*, London: Phaidon Press Limited, 1996.

Wiggington, Michael and Harris, Jude. *Intelligent Skins*. Oxford: Butterworth-Heinemann, 2002.

Willis, Carol. *Building the Empire State: A Rediscovered 1930's Notebook Charts the Construction of the Empire State Building*. New York: W. W. Norton & Company in association with The Skyscraper Museum, 1998.

Zahner, L. William. *Architectural Metals: A Guide to Selection, Specification, and Performance*. New York: John Wiley & Sons, Inc., 1995.

Index

Aalto, A. 32–34, 74, 77
accreditation, wood 89
acid rain 228
advertising 219
aerodynamics 221
aggregates 49–50
air conditioning 100, 179
air gaps 62, 64
aircraft industry 221–22
albedo 232
Alcoa Building 226
aluminum 144–47, 226–27, 232, 240–41; curtain walls 207, 210, 213–14, 216, 221, 226–27
Aluminum Company of America (ALCOA) 226
American Society of Civil Engineers 222
American Society for Testing of Materials (ASTM) 31, 49, 147–48
American Tree Farm 88
Americas 16
Ames Library 46-47, 62
Anasazi 16
anchors 14, 44, 64, 70, 76; cladding 168, 178–80; components 188; curtain walls 216–17, 220–21; environmental forces 229; finishes 238, 240; platform framing 95; spandrel systems 189–90
Ando, T. 60
Annealed Float Glass 129
anodization 232
Antarctic 41
Aqua Tower 126–27
Arab architecture 229
Architectural Ironworks 121
asbestos 102
Aspdin, J. 9
asphalt 99, 102
assemblies of glass 129–31
Astor resort 86
Atelier Jean Prouvé 212
Austro-Hungarian Empire 158
automobiles 93

Badger, D.D. 121, 123
Bagsværd Church 154
Bailey, D.E. 198–99
balloon framing 84–86, 89, 98
barn raising 84, 98
basalt 48
Bauhaus Building 200–202
bearing walls 5-112
Beauvais Cathedral 14, 17
bed joint, definition 36
Bell, I. 86–87
Berlin 233
béton armé 54
Biong Arkitekter 106–8
boat building 81
bond types, brick 37
Boots Factory 181
Boston 62, 132
breakage patterns of glass 129–31
brick categories 30–33, 36–37, 67
Brick Institute of America 31
Brinkman & Van der Vlugt 149–50
Brinkman, J. 149–50
building codes 111, 158, 208–9, 222, 224
Bukhara 11
Bunshaft, G. 132, 213
Burnham and Root 20
Burton, D. 200
Byzantium 9

C-studs 112
Calidarium, Pompeii 9, 11
Candela, F. 54–58
capillary action 71, 167, 169, 228–29
carpentry skills 81, 84, 86, 95, 98, 116
cast iron 120, 200
cast-in-place concrete 49–60
cavity 39, 60–78, 89–90, 112, 124; cladding 178, 180; curtain walls 227–29, 233–34; frame construction 130, 134, 141, 146–48, 153–54; spandrel systems 186–87
cedar shake houses 100

Center for Contemporary Arts, Santiago de Compostela, Galicia, Spain 74, 77–78
ceramic frit 131
certification, wood 88–89
chain-of-custody tracking 89
Chatsworth House 118
Checkpoint Charlie 233
Chicago 20, 31, 84, 126, 156, 181, 207, 218
chimney effect 169 *see also* stack effect
Chrysler Building 203
Church of Christ the Worker, Atlantida 23–29
Church of St Joseph, Le Havre 139
circular arches 38
Cité University 141
cladding 70, 72, 77, 122, 148–49; curtain walls 213, 222, 224, 227; systems 156–96
Cliff Palace 19–20
coated glass 131–32
cold joints 57, 60
Coliseum 9–11
collar joint, definition 36
Cologne Werkbund Exhibition 197
color matching 36
Columbia University 239–40
common bond, definition 37
compression 5, 40, 54, 129, 148, 163, 214
concrete 5–61
concrete block 33–36
concrete characteristics 49–61
concrete masonry units (CMUs) 33–36, 39, 50, 67, 70, 74, 95, 148
condensation 42, 59, 62, 64, 70; cladding 158, 169, 177–80; curtain walls 228; framing 89, 112, 124–25; weather 77; windows 146
conduction 180
Constantinople 9
construction schedules 181, 207, 210
contraction 40
convection 67–68, 71, 227, 233
Convent of La Tourette 54
Le Corbusier 54–55, 141–43
corrosion 70, 125, 145, 228
Cory, Cory and Matsui 181
course, definition 36
coursing 37, 46
crane weight limits 188, 190

creep 125, 148
Crystal Palace 118, 120–22
cullet 129
curing 49–50, 89
curtain walls 167, 180, 197–244; finishes 232–43; fireproofing 207–15; forces/pressures 221–23, 227–31; glass architecture 197; height/movement 216–20; history 198–206; materials 198–206; mullions 227; panelization 224–26; unitized systems 216–20
customization 221–22, 227–28, 232, 238

Davis Brody Bond 240
De Blois, N. 213
De Meuron, P. 48, 57, 103, 131, 141
dead loads 148, 188
Denmark 154
detailing 179, 227–28
dew point 64–65, 70, 112, 124–25, 146
dice 129
Dieste, E. 23
dimensioning 39
Dominus Winery 48–49, 52–53
double glazing 144
double seal systems 176–78
dry glazing 147
Duany/Plater Zyberk 102, 104

earthquakes 11, 67, 136, 148, 200
Eberswalde Technical College Library 57–59
economies of scale 89, 92
efflorescence 39
elevators 121, 156, 181, 207
Empire State Building 203–7
engineered wood products 103, 106
English bond, definition 37
environmental forces 228–32
Enzo-Gutzheit 74
erection tolerances 181, 220–21
Evans, W. 87
evaporation 70
Exeter Library 73–76
expansion 40
Experimental House 32–34
Exterior Insulation and Finish Systems (EIFS) 102

fabricated stone 44, 46
Fair Store 207–8
False Fronts 84

Farnsworth House 122–25
Fédération Nationale du Bâtiments (Fédebat) 149–52
finish siding 100
finishes 232–34
fire escapes 200
fire spread 111–12
fireproofing 20, 156, 207–16
First World War *see* World War I
flame spread *see* fire spread
flanges 83, 210, 212–14
flashing 42, 62, 72–73, 99, 112; cladding 178–80; curtain walls 226–27; windows 136, 148
flat jack arches 41
Flemish bond, definition 37
float glass process 129
flood coats 131
flying buttresses 14
Ford, E. 46
forest management 88–89
Forest Stewardship Council (FSC) 89
Form, formwork 232–34
Foster, N. 162–64, 191–92
foundations 14, 67, 95, 180
frame construction 79–112, 115–54
Frampton, K. 158
Franck, K. 144
frequency 219
fully-tempered glass 129–32, 166

Garden Cities 89
gaskets 136, 146–49, 186, 214, 220–21, 226–27
gauging 40
Gideon, S. 84
Gilded Age 86
Giza pyramids 6
glare 131
glass 128–32, 163, 166, 178, 181; cladding 193; curtain walls 197–200, 210, 213, 216–17, 220; finishes 241–42; pressures/forces 221, 229 227
glass fiber reinforced cement materials (GFRC) 169
Glass House 197
glazier's points/putty 136–37, 144
glazing channels 212
glue laminated wood 103, 106, 109
Gothic arches 14, 116
granite 5, 44, 77

Great Exhibition 118
Great Fire of Chicago *see* Chicago
Great Mosque of Isfahan 11, 14–15
Great Pyramid of Giza 5
Great Wall of China 6
green wood 89
greenhouse construction 198–200
Gropius, W. 94, 198, 200–202
grout 39
GSW Immobilien AG 233
GSW Tower 233–37
gypsum 99–100, 112

Hagia Sofia 9, 11–12
Hallidie Building 200–201
Hancock Tower 134
Harrison & Abramowitz 226
head joint, definition 36
header, definition 36
heat-strengthened glass 129–32
heat-treated glass 129
heating 100
height 216–21
Hennebique, F. 54, 139
Herzog & de Meuron 48, 52–53, 57–59, 103–5, 131, 141, 143
Herzog, J. 48, 57, 103, 131, 141
Highline 23 project 172–76
historical survey 5–33
Home Insurance Building 207
hot spots 222
humidity 41–42, 64, 70, 78, 100, 124, 146
hurricanes 222
Hylar 232

I.M. Pei and Partners 132, 134
Icahn Laboratory 135
igneous stone 44
Incas 16
industrial revolution 115, 198, 232
infill construction 115–54
installation sequencing 180
Institut du Monde Arabe 229–32
insulating glass units (IGUs) 130–32, 144, 146
insulation 40–42, 59–60, 62, 64, 67; cladding 157, 169, 178–80, 186; curtain walls 213, 216, 226, 229, 241; infill systems 124–26; light frame construction 86, 89, 95, 100, 102, 112; weather 70–71, 74; windows 146–47

Les Invalides 16, 67
Isfahan 11, 14–15
Islamic architecture 11
isopropyl alcohol (IPA) 149
Israeli Supreme Court 48–51
Istanbul 11

Jacobs House 90–92
Jefferson, T. 23–24
Jenny, W.L. 207–8
Jewish Community Center Bath House 35–36
John Hancock Tower 132
Johnson Wax Laboratory 183, 185–87
jointing 39, 122, 125, 134, 167; cladding 169–70, 174, 177; curtain walls 222; fillers 180; sizing 148–49
joists 84, 86, 97–98, 106
Jones, F. 103, 105–6
jumbo brick 31

Kaaba 11
Kahn, L. 35, 73–76, 152–53
Karmi and Karmi 48–51
Karmi, R. 48
Karmi-Melamede, A. 48
Katsura Imperial Villa 117, 119
Kelbaugh, D. 67–68
Kelbaugh House 232
Kevin Roche John Dinkeloo and Associates (KRJDA) 50, 53
Kew Gardens 200
Kilta creamer 144
Kursal Cultural Complex 193–96
Kynar 232

Lake Shore Drive Towers 218
laminated safety glass 166
lateral forces 221–24, 228
Laugier, M.-A. 79–80
leakage 167, 186
LEED 89
Lehigh Valley Railroad 181
Lever House 132–33, 209–10
Levitt & Sons 89–90
Levitt Brothers 92
Levittown 92–93
life cycle energy studies 178
light frame construction 79–112
lightweight back up 180
Lillehammer Winter Olympics 106

limestone 5–6, 9, 44, 186, 207
lintels 40
litigation 132
live loads 148, 189, 238
load-bearing construction 5–113; bonds 37; brick terminology 36; cavity walls 61–78; CMU 33–35; dimensioning 39; expansion/contraction 40; flashing 42; insulation 40; jointing 39; light frame 79–113; monolithic bearing walls 5; mortar 36; openings 40; reinforcement 37–38, 40; stone/concrete walls 42–48; structural requirements 23; technique 39; walls 61–112
loads on glass 129
log cabin construction 81–82
Longmeadow stone 46
Loudon, J.C. 199
Louis IX 115
low-emissivity (low-e) coatings 132
lumber 79, 81, 83–84, 89, 103

Machu Picchu 16, 18
McKim, Mead & White 62–64, 86–87
maintenance 72, 144, 181, 210
Los Manantiales Restaurant 54–58
marble 44, 62, 158
Mark, R. 14
Marseilles 141
masonry 5–61; cavity walls 61–62; concrete blocks 33–35; construction technique 39; definition 5; history 5–32; monolithic reinforced 37–38; openings 134
Mason's Bend Community Center 163, 165–66
Mecca 11
Mediterranean Sea 48
Mesa Verde 16, 19–20
metamorphic stone 44
microclimates 100
Middle Ages 14
Mies van der Rohe, L. 121, 123–25, 210–11, 218–19
Ming dynasty 6
minimalism 54
modularization 30, 149, 152, 154, 158
moiré 130
moisture control 70
mold 42, 62, 71, 179
Monadnock Building 20–23, 67

Moneo, R. 193–96, 239–40
monolithic bearing walls 5–62, 66, 70
Morgan Library 62–64
mortar 36, 62, 73
mortgages 90, 92
mortise and tenon connections 83–84, 95, 116–18
movement 216–21
290 Mulberry Street, New York 170
mullions 210, 212–14, 216–20, 227–28, 238, 241
multi-storey projects 126
Muquarnas vaults 14
mushroom columns 183

nails 83–84, 95
Napoleon's Tomb 67
Neil M. Denari Architects 172–75
neoprene 220
Nervi, P.L. 54–55
New York 120–21, 132, 170, 172, 174, 181, 203, 209, 213, 218, 221, 240
Newtons (N) 222
Niels Torp Arkitekter 106–8
non-load-bearing construction 134–48; cladding 156–96; curtain walls 197–243; frame/infill 115–33; insulation 178; precast concrete spandrel 170–75; rainscreen/double seal systems 176–77; vapor barriers 178
Norman brick 30
Northern climates 40
nostalgia 102
Notre Dame de Paris 14, 16–17, 115
Notre Dame de Raincy 138
Nouvel, J. 229–31

oil crisis 67
Olympic Speed Skating Rink 106–8
open joint systems 177
openings 40, 134–44
operation of windows 144–49
Opus Caementicuium 9
Opus Incertum 8
Opus Reticulatum 8–9
Oriented Strand Board (OSB) 99
Ostia 6
oxidation 232
ozone layer 221

Pacific Coast Borax Factory 209–10
Palazetto dello Sport, Rome 55
Palm House, Bicton, UK 198–99
panelization 170, 190, 221, 224–27
Paris 14, 16–17, 60, 67, 115, 125, 138, 149, 229
Parthenon 6
passive solar houses 68
patents 84
Paxton, J. 118, 120
Pei, I.M. 132
Pennsylvania University 152
Pepsi-Cola Building 213
Perret, A. 54, 138–41
Pharaohs 5
Pilkington, A. 129
piping 100
plate glass 128
platform framing 86–112
plumbing 121, 156
Plywood House 103–4
Polk, W. 200
pollution 228
polyvinyl butyral (PVB) 130
Pompeii 9, 11
ponding 169
Portland cement 9, 36, 39, 49, 102
Post Office Savings Bank, Vienna 158–61
post-and-beam 81, 117
post-and-lintel framing 81 see also post-and-beam
pounds per square foot (PSF) 222
Pozzolian stone 9, 11
Prairie Style 32, 90
precast concrete 50–51, 170–76, 187–90, 221, 224
prefabrication 149, 221, 232, 238
pressure equalization 227–28
pressure maps 222, 227
primitive hut construction 79–80
Princeton, NJ 67–68
Princeton University 133, 135
Programme for the Endorsement of Forest Certification (PEFC) 88–89
proto-unitization 149–54
Prouvé, J. 125–26, 149–52, 212, 224
Pueblans 16
pyramids 5–6
Pyrex 186
pyrolitic process 131

Queens Hospital Center 170

rafters 84, 98
rainscreens 68, 72, 176–78, 228, 241–42
rainwater 62–78, 146
Ransome, E. 137, 209–10
re-cladding 191
re-using materials 166
rebar 39–40, 54, 186
Red Rock Canyon 46
Reed, J. 84
reinforced concrete 54, 121, 137–41, 154, 156; cladding 169, 181; curtain walls 200
reinforcement 40, 67, 109
retrofit 131
Richards Medical Laboratories 152–53
Richardson, H.H. 46–47
Ricola-Europe SA 131
Robertson, L.E. 221–23
Robie House, Chicago 31–32
Rock Valley College 108
Roman brick 6, 8, 31–32
Romans 6, 9, 14
rounded arches 41
rowlock, definition 36
Royal Botanical Gardens at Kew 200
25 bis Rue Franklin apartment block 139
Ruff, T. 57–58
rules of thumb 23, 89
running bond, definition 37
runoff 176

safety glazing 129–30, 166
Sainsbury Centre for the Visual Arts 191–92
Sainte Chapelle de Paris 115–16
salvaged materials 166
Sam Mockbee & Rural Studio 163, 165–66
Samanids, Tomb of 11–13
San Francisco 200
sandstone 19, 44, 46
Sauerbruch Hutton 233–37
Säynätsalo Town Hall 74, 77
scaffolding 227
scale models 222
Scheerbart, P. 197
Seagram Building 210–11, 218
sealant sizing 148
seasonal climate cycle 67–68
Second World War *see* World War II

sedimentary stone 44
segmental arches 38
seismic forces 67, 72, 125, 217, 219–21, 224
semi-circular arches 38
shade 40, 100, 126, 129, 131; curtain walls 200, 229, 232, 242; frame/infill constructions 133
Shady Grove Baptist Church 87
shear 44, 67
sheathing 89, 95, 97, 99–100, 102; cladding 180; light frame construction 109, 112
shingle 86, 167
SHoP Architects 170–72
shrinkage 89, 106, 189
silicones 73, 146, 148–49, 180
single seal systems 177–78
site labor 181, 191, 221
site-cast concrete 49–60 *see also* cast-in-place concrete
Siza, A. 74, 77–78
Sjostrom, B. 108–11
Skidmore, Owings and Merrill (SOM) 132–33, 209, 213
skylights 131, 138
skyscrapers 121, 156, 207, 210, 219; curtain walls 226–27; fireproofing 214; pressures/forces 221
slate 44, 141, 144
sliding windows 144
snow 118, 156–58, 197, 228–29
Snow, G.W. 84
Soda Lime Glass 129
soldier, definition 36
Solex 132, 210
spackling 100
spalling 73
spandrels 131–32, 149, 154, 170–76, 180–93
Sports Palace, Rome 54
stack effect 169, 227
stained glass 115–16, 139
standard brick 30
Starlight Theater 108–11
Starrett Investment Corporation 181
Starrett-Lehigh Building 181–84
stave churches 117, 120
stick systems 209–10, 212, 214, 216–19, 238
stone 42–49

stone bearing walls 42–49
Stone House 141, 143
stone types/properties 45
strength of glass 129
stress patterns 132
stretcher, definition 36
structural bonds 37
stucco 102–3, 178
Studio Gang Architects 108, 126–27
studs 84, 86, 89, 95, 98–100, 111–12
subcontractors 89, 179–80
suburbs 92–94, 111, 222
suction maps 222, 227
Supreme Court of Jerusalem 48
surface tension 71, 167, 228
sustainability 88–89
Sustainable Forestry Initiative (SFI) 88
swing windows 144
Swiss Pavilion 141–42
Switzerland 103
Systeme Grille 212
Système Panneaux 224

tableware 144
tar paper 86
tatami 117
Taut, B. 197–98
Taylor, A.D. 84
Taylorism 152, 232
Teema Collection by Arabia 144
temperature gradients 64–65, 70, 144
tempered glass 129, 131–32, 166
tension/tensile strength 5, 36, 39–40, 54, 67; cladding 156; curtain walls 214; frame/infill construction 116, 129, 132–34
textured glass 131
thermal expansion/gradients 122, 124, 144, 148
Thorncrown Chapel 103, 105–6
three-centered arches 38
timber framing 116, 119
tinted glass 132
tolerances see erection tolerances
tooling 39
Torroja, E. 54
torsion 188
Tour Perret 139
town planning 102
Travaux Publics 139
Travertine stone 6

treatment of glass 131–32
Trenton Bath House 35
triple glazing 130, 144
Trombe Wall 232
truss back up 181
Tufa stone 6
turbulence 221
Turner, R. 200

Underwriter's Laboratories (UL) 207
UNESCO chapel, Paris 60
Unité d'Habitation, Marseilles 54–55, 141, 143
unitized systems 216–21, 227–28, 238–42
Usonian houses 90
Utzon, J. 154

Van der Vlugt, L. 149–50
Van Nelle Tobacco Factory 149–50
Vanderbilts 86
vapor barriers 41–42, 59, 70–71, 77, 86; cladding 178–80; curtain walls 214, 228, 241; frame/infill construction 125; light frame construction 89, 100, 112; windows 146
veneer 62, 64, 66–67, 71–73, 77; cladding 158, 178, 180; curtain walls 207, 227; light frame construction 89, 102–3, 112; windows 141
ventilation 100, 144, 229, 234
vernacular architecture 79, 84, 90, 94, 109, 138
vertical transportation 156, 207
Vienna 158–61
Vikingskipet (Viking Ship), Hamar 106–8
Viñoly, R. 133, 135
vinyl wallpaper 71
Virginia, University of 23–24
vision glass 132
Vitruvius 79
vulcanization 147

Wagner, O. 158–61
wattle-and-daub 117, 119
weather 61–78, 86, 100, 156, 221–22; curtain walls 227; pressures/forces 228–29
weathering 100
weep holes/systems 42, 64, 66, 73, 169, 177–80, 227–28

Wesleyan University Creative Arts Center 50–51, 53
Western world 79, 117, 136, 197
wet glazing 149
whitewashing 62, 71, 154
woodland 88
Willis, Faber and Dumas Building 162–64, 166
Willis Polk Architect 201
wind 190, 221–24, 227–28
window washing systems 210
windows 134–49
wood joinery 81

wood-framed house construction 95–100
Worker's Council for Arts 198
World Trade Center (WTC) 221–23
World War II 89, 121, 145, 181, 183, 200, 209, 214, 221
Wright, F.L. 31–32, 90–92, 183, 185–87
wrought iron 200
wythe, definition 36

Yamasaki, M. 221

zoning codes 102, 209